Ecology, Conservation, and Management
of Southeast Asian Rainforests

Ecology, Conservation, and Management
of Southeast Asian Rainforests

EDITED BY RICHARD B. PRIMACK
AND THOMAS E. LOVEJOY

YALE UNIVERSITY PRESS
NEW HAVEN AND LONDON

Sponsored by the Tropical Forest Foundation. The Sarawak Timber Association and Caterpillar, Asia, graciously provided support for including the color illustrations in this book.

Designed by Rebecca Gibb.
Set in Adobe Caslon type by New England Typographic Service, Inc., Bloomfield, Connecticut.
Printed in the United States of America by Book Crafters, Inc., Chelsea, Michigan.

Library of Congress Cataloging-in-Publication Data

Ecology, conservation, and management of Southeast Asian rainforests /
    edited by Richard B. Primack and Thomas E. Lovejoy.
        p.   cm.
    Includes bibliographical references and index.
    ISBN 0-300-06234-6 (alk. paper)
        1. Rain forest conservation—Asia, Southeastern.   2. Rain forests—
Asia, Southeastern—Management.   3. Rain forest ecology—Asia,
Southeastern.   I. Primack, Richard B., 1950–   . II. Lovejoy,
Thomas E.
SD414.A785E36   1995
333.75′0959—dc20   94-45878
                                                            CIP

A catalogue record for this book is available from the British Library.

Printed on recycled paper.
The paper in this book meets the guidelines for permanence and durability of the Committee on Production Guidelines for Book Longevity of the Council on Library Resources.

10   9   8   7   6   5   4   3   2   1

CONTENTS

Acknowledgments vii

Introduction 1
Richard B. Primack

1 Comparing Southeast Asian and Other Tropical Rainforests 5
T. C. Whitmore

PART 1   ECOLOGY AND CONSERVATION

2 Plant Diversity of the Malesian Tropical Rainforest and Its
Phytogeographical and Economic Significance 19
E. Soepadmo

3 Structure, Dynamics, and Management of Rainforests on
Nutrient-Deficient Soils in Sarawak 41
Eberhard F. Bruenig and Hans J. Droste

4 Links Between Vertebrates and the Conservation of Southeast
Asian Rainforests 54
Junaidi Payne

5 Wildlife Responses to Disturbances in Sarawak and Their
Implications for Forest Management 66
Elizabeth L. Bennett and Zainuddin Dahaban

6 Rainforests and Their Soils 87
John Proctor

7 Effects of Selective Logging on Soil Characteristics and
Growth of Planted Dipterocarp Seedlings in Sabah 105
Ruth Nussbaum, Jo Anderson, and Tom Spencer

8 Nursery and Vegetative Propagation Techniques for Genetic
Improvement and Large-Scale Enrichment Planting of
Dipterocarps 116
Pedro H. Moura-Costa

PART 2   POLICY AND MANAGEMENT

9 Long-Term Ecological Research in Indonesia: Achieving
Sustainable Forest Management 129
Herwasono Soedjito and Kuswata Kartawinata

10 Conserving the Reservoirs and Remnants of Tropical Moist
   Forest in the Indo-Pacific Region 140
      Eric Dinerstein, Eric D. Wikramanayake, and Mark Forney

11 Identifying Sites of Global Importance for Conservation:
   The IUCN/WWF Centres of Plant Diversity Project 176
      Stephen D. Davis

12 The Role of Totally Protected Areas in Preserving Biological
   Diversity in Sarawak 204
      *Abang Haji Kassim bin Abang Morshidi and Melvin Terry
      Gumal*

13 The Significance of the Timber Industry in the Economic and
   Social Development of Sarawak 221
      Hamid Bugo

14 Timber Trade, Economics, and Tropical Forest
   Management 241
      Jeffrey R. Vincent

15 Tropical Forests and Climate 263
      R. A. Houghton

   Contributors 291

   Index 293

Color plates follow page 86.

ACKNOWLEDGMENTS

This book developed from a conference organized by the Tropical Forest Foundation with additional financial support from the International Union of Forestry Research Organizations (Special Program for Developing Countries); the United States Department of Agriculture (Forest Service); Caterpillar, Inc.; and the Sarawak Timber Association. Keister Evans of TFF and Barney Chan (STA) provided logistical support throughout the project. Elizabeth Platt was the principal manuscript editor. Boston University provided the facilities for working on the manuscript. Chapters were reviewed by book contributors and outside reviewers, including Peter Ashton, Ian Baillie, Mike Balick, John Beaman, Peter Becker, Dan Binkley, Nick Brokaw, Peter Brosius, Sandra Brown, Richard Condit, Francisco Dallmeier, Marty Fujita, David Gates, Malcolm Gillis, Bruce Haines, Pamela Hall, Ruth Kiew, Ariel Lugo, Alex Moad, Jack Putz, Howard Quigley, Kent Redford, Robert Repetto, Ian Turner, and Kristiina Vogt.

Ecology, Conservation, and Management
of Southeast Asian Rainforests

INTRODUCTION
Richard B. Primack

Public awareness of threats to the world's tropical forests has increased tremendously in the past decade, fed by an avalanche of books, television programs, newspaper articles, and educational materials. The general public is now well aware that tropical forests are fabulously rich in species that are beautiful, interesting, and potentially useful. Moreover, people throughout the world are concerned about the rapid destruction of rainforest resources. With this increased awareness have come intense scrutiny and criticism of the forestry policies of many tropical countries. Some nations have responded to this criticism by examining old policies and, where necessary, revising and reforming them in accordance with relevant new scientific data. In Southeast Asia, however, many governments are stymied by the lack of information on the region's tropical forest species and ecosystem dynamics. Though a boom in scientific activity has accompanied the increased public interest in tropical forests, most international research focuses on American and African tropical forests. Data from these regions are not necessarily applicable to Southeast Asian forests, which have many unique and distinctive features (see Whitmore, this volume; Collins, Sayer, and Whitmore 1991). Further, the research that has been conducted in Southeast Asian rainforests has revealed only a fraction of the forests' biological complexity, with innumerable species and ecological relations still to be discovered and studied. As a result, forestry policies often do not take into account the need for conservation, because the ecological information simply is not available.

In October 1992, a conference was convened in Kuching, Malaysia, to address this very problem. Leading figures from various disciplines came from all over the world to give vital background information and to discuss current research issues. Also present were many government officials, some of whom gave papers to demonstrate the extent of the economic and social factors that they must consider alongside the problem of forest conservation. These diverse viewpoints created an exciting atmosphere in which many areas of concern could be addressed from multiple academic and practical perspectives.

An important goal of the conference was to determine whether Southeast Asian forests are being appropriately managed for sustainable development, defined as "development that meets the needs of the present without compromising the ability of future generations to meet their own needs" (WCED 1987). A related goal was to decide on the measures necessary to improve forest management in instances where sustainable use is not occurring, taking into account four interrelated factors: (1) the amount and type of logging activity in the forests, (2) the effect of logging and land management practices on biological diversity, (3) the impact of the timber industry on traditional rural societies, and (4) the links between logging damage and degradation of such ecosystem functions as maintenance of clean water, atmospheric exchanges of water and gases, erosion control, and flood control. The best methods of reconciling competing claims on Asian rainforests can be determined through an examination of these factors (Soedjito and Kartawinata, this volume). The examination must balance the needs of present and future generations, the needs of people living in tropical Asian countries and people from outside the region, the need for economic development to alleviate poverty and the need to protect endangered species, the need to utilize the best aspects of modern life while preserving the traditions of native people, and the need to utilize forest products to spur economic development without damaging the forest.

The papers presented at the Kuching conference were the foundation for this volume. During the three years since the conference, however, the ideas in most of the papers selected for inclusion have been further refined and developed—in some cases, substantially changed. The chapters have been peer-reviewed, updated, and revised several times before reaching their present form. This book is thus not a compilation of the papers presented at the Kuching meeting but the next logical step in the process that was begun by the conference organizers.

In choosing chapters for this volume, the editors selected a broad range of topics, representing as many different aspects relevant to tropical forest management as was possible within the confines of a single book. In the first chapter, Whitmore introduces many of the background issues common to all tropical forests and highlights the unique nature of Southeast Asian forests. The chapters in Part I, "Ecology and Conservation," analyze topics pertinent to conservation of species and ecosystems. Some authors chose a straightforward discussion of recent fieldwork: Bruenig and Droste, for instance, discuss the different vegetation communities on nutrient-deficient soils in Sarawak, Bennett and Dahaban give results of a study of effects of disturbance on vertebrate communities, and Nussbaum, Anderson, and Spencer examine data on effects of logging on tree seedlings. Other authors analyze more general aspects of tropical forest management: Payne, for example, outlines the links between forest conservation and vertebrate communities throughout the region, Proctor discusses rainforest soils in Southeast Asia, and Moura-Costa presents his findings on trials of different dipterocarp species for genetic improvement of timber stock.

In Part II, "Policy and Management," the chapters chiefly examine the conflicting needs of forestry officials: on one hand, to conserve sufficient forest to maintain healthy populations of plant and animal species, and on the other, to use forest resources to support the needs of the local population. Chapters by Soedjito and Kartawinata; Dinerstein, Wikramanayake, and Forney; Davis; and Abang Morshidi and Gumal tackle different aspects of conservation policy from a research perspective. One particular theme is found in all four chapters: more data are desperately needed to establish conservation priorities, but in the meantime the threats are too great to allow valuable species to go unprotected. These chapters are in contrast to the chapters by Bugo and Vincent, which offer the human perspective on conservation. As both authors point out, many of the Southeast Asian countries that are richest in natural resources are also underdeveloped and strapped for capital. These chapters raise a crucial question: How can conservation be reconciled with growing human pressure for income when, in many instances, local people are dependent on forest resources for subsistence? A final chapter by Houghton brings the discussion full circle: Houghton emphasizes the immediate need for solutions to these problems by highlighting their global impact, demonstrating the importance of rapid action to conserve tropical forests in Southeast Asia and throughout the world.

Many chapters focus on Sarawak, a state of Malaysia on the northwest coast of Borneo. Sarawak has emerged as the world's leading exporter of tropical logs, allowing it to enter a period of rapid economic growth (ITTO 1990; Primack 1991; Primack and Hall 1992). But because Sarawak is one of the world's centers of diversity of plant and animal species and is home to many indigenous people practicing traditional ways of life, its timber policy has been targeted for criticism by international conservation and human rights groups. Sarawak has a unique combination of attributes: valuable forests, considerable oil and natural gas wealth, a low density of people in a largely forested land-scape, a well-run Forest Department, and a sophisticated state government. The state is a key test case of the concept of sustainable development of the tropical rainforest; for this reason, in this volume Sarawak is often treated as a microcosm of Southeast Asian forests.

Yet Sarawak's situation is not representative of conditions in other parts of Southeast Asia, where threats to tropical forests are greater or more im-mediate, while attempts to halt destructive activities are fewer. If conservation efforts are going to succeed anywhere in Southeast Asia, they should do so in Sarawak. If, in contrast, conservation efforts fail in a place with all of Sarawak's advantages, they may not work anywhere in the region.

REFERENCES

Collins, N. M., J. A. Sayer, and T. C. Whitmore, editors. 1991. *The Conservation Atlas of Tropical Forests: Asia and the Pacific.* Macmillan, London.

ITTO (International Tropical Timber Organization). 1990. The promotion of sustainable forest management: A case study in Sarawak, Malaysia. Report.

Primack, R. B. 1991. Logging, conservation and native rights in Sarawak forest. *Cons. Biol.* 5: 126–130.

Primack, R. B., and P. Hall. 1992. Biodiversity and forest change in Malaysian Borneo. *BioSci.* 42: 829–837.

Soedjito, H., and K. Kartawinata. This volume. Long-term ecological research in Indonesia: Achieving sustainable forest management.

Whitmore, T. C. This volume. Comparing Southeast Asian and other tropical rainforests.

WCED (World Commission on Environment and Development). 1987. *Our Common Future.* Oxford University Press, Oxford.

I   COMPARING SOUTHEAST ASIAN AND OTHER
TROPICAL RAINFORESTS
T. C. Whitmore

INTRODUCTION: THE WORLD'S TROPICAL RAINFORESTS
The world's tropical rainforests occur in three main blocks (fig. 1.1). Tropical America has about half the total, with potential cover (before artificial attrition) of roughly 4 million km², mostly in the Amazon and Orinoco basins. Southeast Asia comes second, with approximately 2.5 million km², centered on the Malay archipelago but extending into continental Southeast Asia as far as Sri Lanka and India, and into Queensland (10,000 km²), Melanesia, and Polynesia. The heartland is the region known to botanists as Malesia. It extends from the Kra Isthmus (the narrow part of the Malay Peninsula) through the archipelago to the Torres Strait (between Australia and New Guinea) and the Bismarck Archipelago (east of New Guinea). The third and smallest rainforest region is that of Africa, approximately 1.8 million km², mainly the Zaire basin of central Africa but extending westward along the shores of the Gulf of Guinea.

Southeast Asia is much more mountainous than Africa or South America and has extensive young, eroding landscapes. It also has substantial areas of volcanic rocks and soils (Richter and Babbar 1991).

Climate

Tropical rainforests are found in climates that have no dry months (rainfall 60 mm), or only a few dry months (see fig. 1.1). The tropical evergreen rainforest formation occurs in places with no dry season; usually all months are wet (rainfall 100 mm). Tropical semi-evergreen rainforest forms where there is a short dry season. In still drier climates rainforests are replaced by tropical seasonal forests, called monsoon forests in Asia. Tropical semi-evergreen rainforest differs from the evergreen formation mainly in that some of the canopy-top and emergent species are deciduous. It also differs in its species composition (though many genera are the same as in evergreen rainforest), and it sometimes has stands in which one or a few species of trees are either dominant or very common.

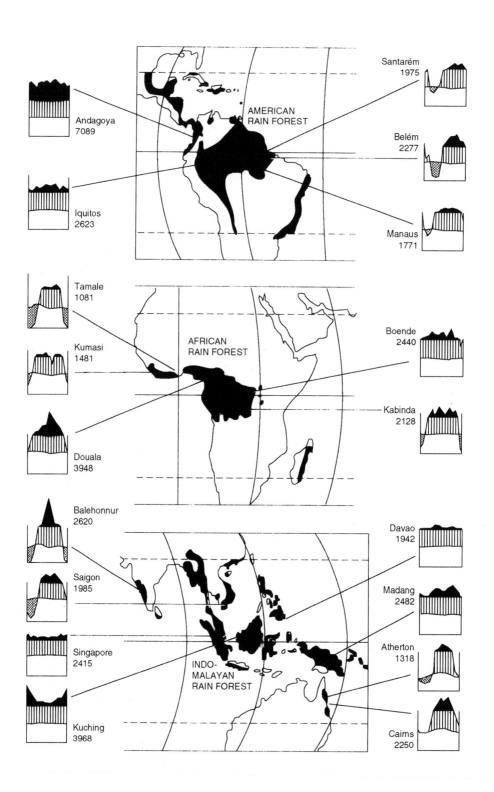

Andagoya
7089

Iquitos
2623

Santarém
1975

Belém
2277

Manaus
1771

AMERICAN
RAIN FOREST

Tamale
1081

Kumasi
1481

Douala
3948

AFRICAN
RAIN FOREST

Boende
2440

Kabinda
2128

Balehonnur
2620

Saigon
1985

Singapore
2415

Kuching
3968

Davao
1942

Madang
2482

Atherton
1318

Cairns
2250

INDO-
MALAYAN
RAIN FOREST

The whole of the Asian tropics has evergreen rainforest except for the fringes and outliers, whereas Africa has almost entirely the semi-evergreen formation. The lower (eastern) Amazon forests are semi-evergreen, and parts have three or four dry months; the western, upper Amazon, closer to the Andes, has evergreen rainforest (Baur 1968).

Floristic Richness

All tropical rainforests are fabulously rich in both plant and animal species (Whitmore 1984, 1990), and they appear to have more species than any other ecosystem on earth. Consider just the trees: The whole of Europe north of the Alps and west of the Commonwealth of Independent States has 50 native tree species, and all of eastern North America has 171. By comparison, small sample plots of tropical rainforest commonly have 100 species (and many have more than 200) of trees at least 10 cm in diameter at breast height (dbh) per ha (fig. 1.2).

Journalists and science writers often state that the Amazon rainforests are richest in numbers of species. This assertion is misleading. The reason that the Amazon rainforests have so many species is that they are by far the most extensive. The island of Borneo, for example, would fit several times over into the Amazon basin. It is necessary to allow for area to obtain a meaningful measure of species richness. One measure that is widely used to quantify species richness is the number of tree species with individuals greater than 10 cm dbh on small forest research plots. Using this measure, in general more tree species are found in one place in Southeast Asia than in the Neotropical (American) rainforests, which are themselves richer in species than plots in Africa (see fig. 1.2). Two exceedingly rich plots, however, have been enumerated at Yanamomo in Peru, near the Andes (Gentry 1988a). One of these, with 283 species per ha, is the richest forest yet described; every second tree is a different species. The poorest rainforests are in West Africa, with just 23 species per ha in one Nigerian

FIGURE 1.1 Tropical rainforest distribution and climate (fig. 2.1 in Whitmore 1990). The *Klimadiagrammen* of yearly rainfall patterns follow the conventions of their inventors (Walter and Lieth 1967). The *x*-axis represents the annual climate pattern, running from January to December for the Northern Hemisphere and July to June for the Southern Hemisphere. The *y*-axis shows both temperature and precipatation. The horizontal line in each figure is the monthly average temperatures; in all sites except Atherton this temperature is high and fairly uniform throughout the year. The second and varying line shows monthly rainfall; dotted regions are relatively dry months, vertically shaded and black areas are humid months, and black areas are the wettest months. For example, Saigon has a dry season early in the year and abundant rainfall for the remainder.

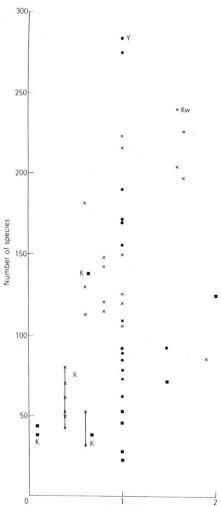

FIGURE 1.2 Species richness among trees of 10 cm in diameter and larger on small plots in tropical lowland rainforest: • = America (Y = Yanamomo), X = Asian tropics (Kw = Kalimantan Wanariset), and ■ = Africa (K = Korup) (fig. 2.27 in Whitmore 1990).

sample, though this is still a high number by temperate standards. Other forests in the upper Amazon region are likely as rich as the Yanamomo forest plot. Indeed, a forest plot at Cuyabeno in Amazonian Ecuador has recently been discovered to have 307 species per ha (H. Balslev, personal communication, July 1993).

Floristic Composition

There are many floristic similarities among the American, African, and Asian rainforests. Gentry (1988b, 1992) has described these similarities in some detail (1988b, 1992). For example, certain tree families, such as the Caesalpinaceae (pea family), Sapotaceae (sapodilla family), Annonaceae (soursop family), Ebenaceae (ebony family), and Myristicaceae (nutmeg family) are commonly found in all major tropical forest regions. There are also important floristic differences, however. For example, the Neotropical rainforests have abundant trees in the Lecythidaceae (Brazil nut) family, with 11 genera and 120 species, and in the Vochysiaceae, a small family with several timber species. Southeast Asian forests have numerous conifers (there are only two conifer species in the Neotropical forests and one in the African rainforests) and several hundred species of the tree genus *Calophyllum,* known as bintangor, which has only four species in the Neotropics and none in continental Africa. One important feature of Southeast Asian rainforests is the abundant climbing palms (tribe Calameae) called rattans. Africa has only a few species of *Calamus,* and America has none. The other important Southeast Asian tree family, the Dipterocarpaceae, is abundant and includes numerous species of timber trees.

The Major Lowland Forest Formations

The rainforests discussed so far occur on dry land, with soils of oxisols and ultisols. In addition, there are rainforests on nutrient-poor soils known as spodosols and two important swampland forest formations, described in greater detail by Bruenig and Droste in Chapter 3.

Heath Forest

This formation has a distinctive structure and physiognomy, consisting of a high density of slender trees, often relatively short in height (Whitmore 1984). Heath forest occurs on spodosols (podzolized siliceous sands) in all continents. These forests are extensive in Borneo, especially in South Kalimantan, and occur elsewhere in Malesia (Whitmore 1989) and extensively in South America (fig. 2.4 in Prance 1987). There are also a few small areas on the coast of West Africa. Heath forest sites are easily degraded, either by opening of the canopy, which allows full solar radiation to reach the surface and oxidize the superficial humus, or by heavy machinery. Open shrubland or bare white sand develops after this heavy damage. Most heath forests have little commercial timber. They can be used for sustainable timber production by a light selection felling, as at Menchali Forest Reserve, Pahang, Malaysia. There were formerly extensive bindang (*Agathis borneenis*) forests of high stocking in South Kalimantan, near Sampit. These were heavily exploited, and there has been poor regeneration.

Peatswamp Forest

This formation is extensive in Malesia (fig. 1.3) but very limited in Africa and the Neotropics (one example is the pegasse swamps of Guyana). Most Southeast Asian peatswamp forests have high volumes of commercially

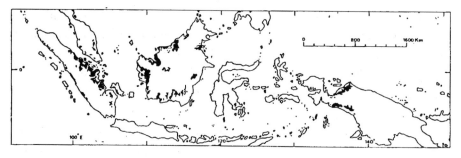

FIGURE 1.3   Peatswamp forests of Malesia (from Whitmore 1989). Only very small areas of this formation are found in Africa and the Neotropics.

valuable timber, including ramin (*Gonystylus bancanus*), the premier timber for moldings. The small areas of peatswamp forest in America and Africa are not commercially important. Where the peat is deeper than about 2 m, peatlands are best left under forest. When the rainforest is cleared from such areas and the surface has dried out, the peat easily burns, shrinks, and is very acid and infertile with extremely low amounts of plant mineral nutrients. Plantation trees on peat are unstable and prone to windthrow. On deep peat, sustainable agriculture or plantation forestry is impossible.

Mangrove Forest

Coastal mangrove forests are the other commercially valuable wetland rainforest formation in Southeast Asia. They occur throughout the region in areas with salty or brackish water. The second largest mangrove forest in the world occurs in western Irian Jaya, exceeded in area only by the Sunderbans forest of Bangladesh and India. The Asian mangrove forests are the most species-rich mangroves in the world. They are under heavy pressure from exploitation for wood chips, conversion to prawn and fish ponds, and filling-in to provide building land. Most Philippine and Singapore mangrove forests have been lost in one of the last two ways. There has been a recent surge of interest in mangrove forests, and numerous reviews (summarized in Ong 1987) have synthesized this large body of research.

DIPTEROCARP RAINFOREST

The Dipterocarpaceae are a family of numerous tree species centered in Southeast Asia. Species in the family have resin canals in their wood. The fruit of most species consists of a nut enclosed by some or all of the flower sepals enlarged as big wings. Dipterocarp species give those Southeast Asian rainforests in which they occur special features. Most rainforest dipterocarps are huge trees—up to 40–50 m or more in height, and up to 2–3 m in diameter. They form giant forests, equaled in stature only by the temperate coniferous rainforests of the northwestern seaboard of North America in Oregon, Washington, and British Columbia. The greatest concentrations of dipterocarp species are found in Sumatra, Malaya, and Borneo (Ashton 1982). Borneo alone has 9 genera and 287 species of dipterocarps. No other tropical rainforests anywhere in the world are so dominated by a single family of trees. There is a dramatic decrease in the importance of dipterocarps east of Borneo. Sulawesi has only 2 genera and 4 species of dipterocarps. The family extends eastward as far as the Louisiade Archipelago and is present in continental Asia north of Malesia, but with few genera and species. Within western Malesia the proportion of the big trees in the forest that are dipterocarps increases eastward from Sumatra through Malaya to Borneo and the southern Philippines, where, as in eastern Sabah, almost all the canopy-top and emergent trees belong to this one family (Whitmore 1984).

Nearly all dipterocarp species have useful timber. Commercially they can be grouped into Light Hardwoods (such as *Shorea*), Medium Hardwoods (such as *Dryobalanops*), and Heavy Hardwoods (such as *Hopea, Shorea* section *Shorea, Vatica*). The stocking in eastern Sabah is often 80–100 m³/ha (Deramakot Forest Reserve had 107 m³/ha on average), and in the southern Philippines more than 300 m³/ha has been recorded. Such heavily stocked forests are virtually destroyed if all the timber is removed. Elsewhere dipterocarp volumes of 30 m³/ha are more common, comparable to the commercial stocking of many African or American rainforests.

East of Borneo, where dipterocarps are uncommon and localized, the forests have lower stature and stocking; common commercial trees include *Calophyllum* (bintangor), *Intsia* (merbau), *Pometia* (kasai), and *Pterocarpus*.

Gregarious Flowering and Fruiting

In addition to their abundance and species-richness, western Malesian dipterocarps have another unique property, namely, gregarious flowering and fruiting two or three times per decade. Across a large area, many, if not most, dipterocarp species will flower within a few weeks of one another, even if no flowering has occurred for several years. Such episodes are usually followed by the appearance of many species of seedlings, at a density of several million plants per hectare. The seedlings typically die away over several years and have for the most part disappeared by time of the next mass flowering. If, however, a canopy gap opens because a tree dies or falls over, seedlings below the gap are exposed to higher light levels and start to grow rapidly. Most climax tree species in the world's forests have this property to respond to canopy gaps. Dipterocarps, however, are unusual in their development of dense seedling populations as a consequence of mass fruiting. Elsewhere in the tropics, some Vochysiaceae of South America fruit gregariously, and periodic heavy fruiting is also a feature of some northern hemisphere trees, such as beeches (*Fagus*), oaks (*Quercus*), and elms (*Ulmus*), but no other tropical tree family behaves in this manner.

Silviculture

From the forester's point of view, the periodic development of huge seedling populations is a very useful characteristic. Dipterocarp rainforests are a silviculturist's dream because they are relatively easy to regenerate. For a new crop of dipterocarp trees to grow at a site that has been logged, one simply needs to fell while there are adequate seedlings on the ground, to control the degree of canopy opening in order to keep out secondary (pioneer) tree species and climbers, and, most important, to keep logging damage sufficiently low.

Foresters in the Southeast Asian region have solved the technical problems and have devised managerial and silvicultural rules for the successful regeneration of dipterocarp rainforests following logging (Baur 1968; Whitmore 1984). Enforcement of these rules is essential, but it is not commonly achieved. All too often, unskilled or unsupervised logging teams cause excessive and unnecessary damage when felling, which hinders the recovery of high forest.

TROPICAL HARDWOOD INTERNATIONAL TRADE

Most internationally traded tropical hardwood timber comes from Southeast Asia (fig. 1.4). West Africa was relatively more important in the decades after World War II, but its export trade has declined since then. Central Africa, in

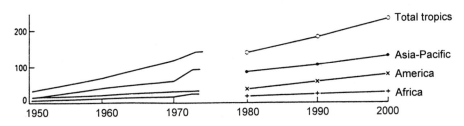

FIGURE 1.4   Actual (1950–1973) and estimated (1980–2000) tropical hardwood production in millions of m³ (fig. 10.21 in Whitmore 1990). FAO (1993) figures confirm that these estimated curves closely approximate actual production until 1990.

particular the Zaire basin, remains scarcely exploited for logistic reasons (Sayer, Harcourt, and Collins 1992). South America has relatively little export trade in timber, but internal markets are large and growing. In Brazil during the past decade, for example, a large movement of timber has begun from the lower Amazon rainforests southward along the paved highway via Brasilia to markets in the densely populated southern part of Brazil. A new development is also taking place in Guyana, where in 1991 a logging license was given for 800,000 ha at Barama to a Taiwanese and Malaysian consortium. The Barama area contains a substantial volume of Light Hardwoods (such as *Catostemma*) that are being exported, partly as plywood.

Tropical hardwood production has increased with time (fig. 1.4). Within Malesia there has been a progressive shift in principal exporters of logs from the Philippines to Indonesia to Sabah and now to Sarawak (fig. 1.5).

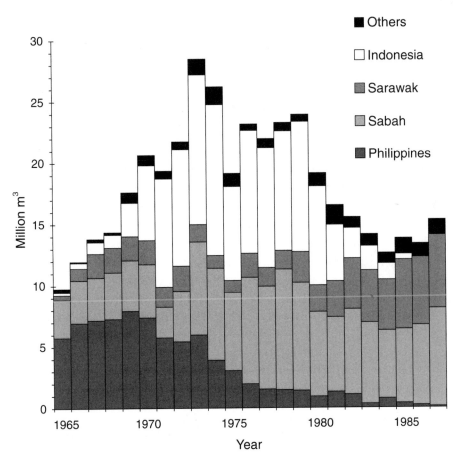

FIGURE 1.5   Log production in Malesia (from Nectoux and Kuroda 1989).

Currently only Sabah, Sarawak, and Papua New Guinea allow the export of unprocessed logs without heavy tariffs.

The Dominance of Southeast Asia in the Hardwood Trade

Southeast Asian rainforests dominate the international tropical hardwood trade for several reasons. First, the forests have a high concentration of commercial timber species, many of them dipterocarps. Second, a high proportion of the species has Light Hardwood timber, for which market demand is the strongest. This proportion is not found in areas such as the lower Amazon. Timber surveys in the lower Amazon by the United Nations Food and Agriculture Office have discovered that very dense timbers are by far the most frequent over a surveyed area of 63,000 km^2 of this region. There is little export demand for these Heavy Hardwoods. Thus, even though some Amazonian forests have 70 m^3/ha of timber, it is difficult to sell this timber internationally because of its high density (Whitmore and Silva 1990). Demerara Timbers Limited of Guyana is trying to develop markets for the Heavy Hardwoods that predominate in its concession area.

Southeast Asian rainforests are extremely rich in big timber tree species as well as small ones. Timber users group species into classes for particular usages. Southeast Asian timber, mainly from dipterocarps, can be grouped for marketing purposes into only a handful of these end-use classes, guaranteeing a continuing supply of large volumes of timber. Thus a strong market has developed. Neither African nor South American species can be so successfully grouped. Most African and South American tree species must be treated differently by the users. It is therefore difficult to guarantee a continuing supply of any particular timber. From the users' point of view, Asian rainforests produce large quantities of a few sorts of timber, whereas African and American ones produce small parcels of many different sorts.

Southeast Asian rainforests are not the same as tropical forests in other parts of the world. Because of the dominance of the Dipterocarpaceae in western Malesia, these forests have special advantages in forest management and timber marketing, and the dipterocarps' gregarious fruiting makes their silviculture easily managed. Rapid felling of trees in many countries, however, has led to the overharvesting of forest and the local exhaustion of timber supplies.

REFERENCES

Ashton, P. S. 1982. Dipterocarpaceae. *Flora Malesiana* Ser. I, 9: 237–552.

Baur, G. N. 1968. *The Ecological Basis of Rain Forest Management.* Forestry Commission of New South Wales, Sydney.

FAO (Food and Agriculture Organization). 1993. Summary of the final report of the forest resources assessment 1990 for the tropical world. Paper presented at the eleventh session of the Committee on Forestry. FAO, Rome.

Gentry, A. H. 1988a. Changes in plant community diversity and floristic composition on environmental and geographical gradients. *Ann. Missouri Bot. Gar.* 75: 1–34.

———. 1988b. Tree species richness of upper Amazonian forests. *Proc. Nat. Acad. Sci. USA* 85: 156–159.

———. 1992. Tropical forest biodiversity: Distributional patterns and their conservational significance. *Oikos* 63: 19–28.

Nectoux, F., and Y. Kuroda. 1989. *Timber from the South Seas.* WWF International, Gland, Switzerland.

Ong, J. E. 1987. Review of P. B. Tomlinson's *The Botany of Mangroves. Trends in Ecology and Evolution* 2: 111.

Prance, G. T. 1987. Vegetation. In *Biogeography and Quaternary History in Tropical America.* T. C. Whitmore and G. T. Prance, editors. Clarendon Press, Oxford.

Richter, D. D., and L. I. Babbar. 1991. Soil diversity in the tropics. *Adv. Ecol. Res.* 21: 316–390.

Sayer, J. A., C. S. Harcourt, and N. M. Collins, editors. 1992. *The Conservation Atlas of Tropical Forests: Africa.* IUCN, Gland, Switzerland, and Macmillan, Basingstoke, England.

Walter, H., and H. Lieth. 1967. *Klimadiagramm Weltatlas.* Fischer, Jena, Germany.

Whitmore, T. C. 1984. *Tropical Rain Forests of the Far East.* Clarendon Press, Oxford.

———. 1989. Southeast Asian tropical forests. In *Ecosystems of the World* 14. B. H. Lieth and M. J. A. Werger, editors. Elsevier, Amsterdam.

———. 1990. *Introduction to Tropical Rain Forests.* Clarendon Press, Oxford.

Whitmore, T. C., and J. N. M. Silva. 1990. Brazilian rain forest timbers are mostly very dense. *Commonw. For. Rev.* 69(1):87–90.

PART I Ecology and Conservation

2   PLANT DIVERSITY OF THE MALESIAN TROPICAL
RAINFOREST AND ITS PHYTOGEOGRAPHICAL
AND ECONOMIC SIGNIFICANCE
E. Soepadmo

The tropical rainforests of Southeast Asia, particularly in Borneo and Papua New Guinea, are widely acknowledged as among the most species-rich and complex terrestrial ecosystems in the world. In the so-called Malesian phytogeographical zone, extending from Peninsular Thailand (the Kra Isthmus) in the northwest to Papua New Guinea and adjacent islands in the southeast and occupying a total land area of approximately 3 million km^2, more than 40,000 species of vascular plants have been recorded. Of these tropical species, slightly more than 36,000 are flowering plants distributed in 266 families and 3,075 genera, whereas 3,600 are ferns and allies, representing 35 families and 164 genera, and 87 species are conifers, which belong to 5 families and 12 genera (de Laubenfels 1988; Roos 1992). By comparison, all of Europe, a land area of about 9 million km^2, harbors about 11,500 species of vascular plants.

The high species diversity found in the natural vegetation of Malesia is, according to many authors (Ashton 1984; Ashton, Givnish, and Appanah 1988; Beaman and Beaman 1990; Beaman et al. 1985; Flenley 1979, 1984; R. J. Johns 1986, 1989, 1990; Moreley and Flenley 1987; Newsome and Flenley 1988; van Steenis 1969, 1979; Whitmore 1981, 1987), attributed partly to the prevailing moist, warm climatic conditions and the availability of diverse types of microhabitats resulting from past and recent geological history of the region, and partly to the ability of the indigenous and introduced plant species to adapt, compete, evolve, and invade newly formed habitats. In addition, the existence of different reproductive systems among sympatric plants of different taxa, and the frequent occurrence of artificial or natural disturbances, may influence the species richness of a particular region. The complex structure and high species diversity of plant communities in tropical rainforests in Malesia also have a direct bearing on the ecological roles of the forests in maintaining and stabilizing such environmental systems as hydrological and nutrient cycles, food chains and energy flows, heat and energy balance, and the carbon sink or storage system. The elimination of these rainforests may have a pronounced effect on local, regional, and global climate (Brown and Lugo 1982; Deteriler and Hall 1988; Palm, Houghton, and Melillo 1986; Soepadmo 1983, 1993; Vitousek and Matson 1992). In addition, the complexity of the natural forests in the region provides a diversity of habitats for wildlife.

THE DIVERSITY OF LIFE FORMS

The structural complexity of tropical rainforests in Malesia is due largely to the presence of diverse plant species in particular habitats. In the lowland and hill dipterocarp forests, for instance, many trees belong to the Dipterocarpaceae (such as *Dryobalanops aromatica*, *Shorea amplexicaulis*, and *S. laevis*) and Leguminosae (such as *Koompassia exelsa* and *K. malaccensis*), which may reach a total height of up to 84 m and a diameter above the buttresses of up to 460 cm (Yamakura et al. 1986). In contrast, a number of terrestrial orchids (such as *Corybas* spp.), aquatic plants (such as *Indotristicha*), and saprophytic flowering plants (such as *Thismia* spp.) attain a maximum size of only a few centimeters. Between these extremes are many small or medium-sized trees and shrubs that occupy the main and understory canopy layers of the forests. Examples include many species of the Annonaceae, Apocynaceae, Burseraceae, Clusiaceae, Euphorbiaceae, Lauraceae, Leguminosae, Meliaceae, Moraceae, Myrtaceae, Rubiaceae, Sapindaceae, Sapotaceae, and several other families. The forests also contain many species of climbers, including rattans and members of the families Apocynaceae, Asclepiadaceae, Connaraceae, Leguminosae, Menispermaceae, Rubiaceae, and Vitaceae, as well as epiphytes (most ferns and orchids, for example), stem and branch parasites (holoparasites such as *Rafflesia* spp. or hemiparasites such as *Dendrophthoe* and *Macrosolene* spp.), stranglers (*Ficus* spp.), terrestrial herbaceous plants, and many others. Depending on the taxa, these plants possess different growth habits and architectures, and modes of nutrition and reproduction, which enable them to compete and live together in equilibrium with the prevailing environmental conditions of a particular forest ecosystem.

Species Diversity

The tropical rainforest of Malesia is one of the most species-rich terrestrial ecosystems, comparable to if not richer than tropical rainforests elsewhere in the world. Many speciose families dominate the vegetation, and detailed ecological studies carried out in particular localities confirm the region's high species diversity. According to Roos (1992), the entire Malesian region has at least 16 families of flowering plants with more than 500 species each

Table 2.1 Large families in the flora of Malesian rainforests

Family	Genera (World/Malesia)	Species (World/Malesia)	% Species found in Malesia
Acanthaceae	357 / 37	4,350 /625	14
Annonaceae	128 / 49	2,050 /875	43
Araceae	106 / 35	2,950 / 775	27
Arecaceae (Palmae)	207 / 53	2,675 / 975	34
Ericaceae	103 / 12	3,350 / 740	22
Euphorbiaceae	326 / 91	7,750 / 1,000	13
Gesneriaceae	146 / 26	2,400 / 900	38
Lauraceae	45 / 22	2,200 / 700	32
Melastomataceae	215 / 39	4,750 / 1,000	21
Moraceae	48 / 14	1,200 / 575	48
Myrsinaceae	39 / 18	1,250 / 500	40
Myrtaceae	121 / 34	3,850 / 1,600	42
Orchidaceae	795 / 217	17,500 / 6,500	37
Rubiaceae	637 / 134	10,700 / 2,000	19
Scrophulariaceae	222 / 31	4,500 / 500	11
Zingiberaceae	53 / 26	1,200 / 575	48

Sources: Data from Mabberley 1987; van Steenis 1987; and Roos 1992.

(table 2.1). Of these, the Orchidaceae is the largest, with about 6,500 species. In addition, 20–25 families have more than 100 but fewer than 500 species each. As van Steenis (1987) and van Welzen (1992) have shown, approximately 40% of the genera and species of the Malesian flora are either endemic to one island or confined to a group of islands in the region, particularly in the northern parts of Borneo (Sarawak, Brunei, Sabah, and North Kalimantan) and Papua New Guinea, which experienced an active and dynamic geological history.

If trees with a diameter at breast height (dbh) of 10 cm or larger are considered, the 50 ha plot at Pasoh Forest Reserve contains 660 species in 244 genera and 67 families. The species diversity and density of trees per hectare at Pasoh Forest Reserve are comparable to those of other lowland rainforests in Malesia, and they exceed those of much larger plots in other tropical regions, such as the 36,000 km² forest at Corocovado National Park, Costa Rica, which has 238 species, or the 170 ha plot at Rio Palenque, Ecuador, which has 154 species (Croat 1978; Gentry and Dodson 1987; Hartshorn 1983; Kochummen, LaFrankie, and Manokaran 1990). Detailed ecological studies on the diversity and distribution patterns of the flora of Mt. Kinabalu in Sabah by Beaman and

Beaman (1990) have recorded about 4,000 species of vascular plants, representing 180 families and 950 genera, within an area of about 700 km². By comparison, only 1,443 species are known in Great Britain, which has a total land area of about 244,000 km², and 1,996 species in New Zealand, with a total land area of around 268,000 km². Likewise, Chin (1977, 1979, 1983a, 1983b), in his studies of the flora of limestone hills of Peninsular Malaysia covering a total area of 260 km², encountered 1,216 species of vascular plants representing 124 families and 582 genera. Anderson (1963) recorded 418 species of vascular plants in 250 genera and 79 families in the peatswamp forest of Sarawak and Brunei, covering a total area of 15,488 km². Of these, 382 species are flowering plants, 33 are ferns and allies, and 3 are gymnosperms. The 10 most speciose families encountered at Pasoh Forest Reserve are listed in table 2.2.

Tree Density

As a consequence of high species diversity, the tropical rainforest of Malesia is composed largely of rare or endemic tree species with a very low density. Excepting those forest types in specialized habitats (such as mangroves and the *Shorea albida* phasic community of Sarawak's peatswamps), no single species or family dominates. The rarity or low density of conspecific trees is even more prominent if we consider only trees with a dbh of 30 cm or greater. Studies by Poore (1968) and Ho, Newbery, and Poore (1987) at the Jengka Forest Reserve found that each of the approximately 38% of the tree species reaching a dbh of 30 cm or larger was represented by a single mature individual. Even the most common species on the plots account for only a small percentage of the total number of trees (Primack and Hall 1991).

PHYTOGEOGRAPHIC RELATIONS OF THE MALESIAN FLORA

Phytogeographically, the Malesian region can be considered the meeting point of at least two major world floristic zones, continental Asia and Australia. Whereas the landmass of Asia has its origin in the Laurasia fragment of the supercontinent Pangaea, Australia has been derived from the Gondwana fragment. As a consequence of geological events of the past 180 million years, the geologically recent geomorphology, and the warm, moist climate, unique and varied habitats conducive for the development of the species-rich flora in the Malesian region have been established. In his analysis of the generic flora of seed-bearing plants of the region, van Steenis (1950) recognized five major categories of components in the Malesian flora:

1. Widespread genera without a distinct center of distribution in the Paleotropics (Asia, Australia, and Malesia). Many of the families

Table 2.2 The ten most speciose families of plants 1.0 cm dbh or greater in the 50-ha plot of Pasoh Forest Reserve

Rank	Family	Species (Peninsular Malaysia / Pasoh)[a]
1	Euphorbiaceae	371 / 87
2	Myrtaceae	210 / 50
3	Lauraceae	213 / 48
4	Rubiaceae	450 / 46
5	Annonaceae	198 / 44
6	Meliaceae	91 / 43
7	Anacardiaceae	74 / 32
8	Clusiaceae	121 / 31
9	Myristicaceae	53 / 31
10	Dipterocarpaceae	155 / 30

Source: Data from Kochummen, LaFrankie, and Manokaran 1990.
[a]Values are the number of species in each family in Peninsular Malaysia followed by the number of species in the 50-ha plot at Pasoh Forest Reserve.

listed in table 2.1, except the Zingiberaceae, are examples of this category.

2. Genera with a center of distribution on the Asian continent (absent or poorly represented in Australia). *Rhododendron* (Ericaceae) and *Castanopsis* and *Lithocarpus* (Fagaceae) are good examples.

3. Genera with the center of distribution in the Malesian region. Examples are *Artocarpus* (fig. 2.1), *Calamus, Nepenthes,* and *Rafflesia,* as well as many genera in the Dipterocarpaceae and Bombacaceae.

4. Genera with the center of development in Australia (absent or very poorly represented in continental Asia), including many genera of the Araucariaceae, Cyperaceae (*Gahnia*), Loranthaceae, Myrtaceae, Pittosporaceae, Proteaceae, and Weinmanniaceae.

5. Genera with the center of development in the Pacific-Subantarctic region, such as *Hebe* (Scrophulariaceae), *Nothofagus* (Fagaceae), and *Oreobolus* (Cyperaceae).

Of 218 endemic genera involving about 386 species, 61 genera are confined to Borneo, 5 to the Celebes, 5 to Java, 1 to the Lesser Sunda Islands, 4 to the Moluccas, 24 to Peninsular Malaysia, 23 to the Philippines, 82 to Papua New Guinea, and 13 to Sumatra. At the familial level, however, only one family, Scyphostegiaceae, with a monotypic genus *Scyphostegia,* is confined to Borneo.

FIGURE 2.1 Chempedak (*Artocarpus integer*) is an important wild fruit tree that is also extensively cultivated in villages. The large fruits are produced directly from the trunk and main branches. Photograph by R. Primack.

Despite the close affinity between the Malesian flora and those of continental Asia and Australia, the Malesian flora has many distinct elements. In the modern flora many genera of seed-bearing plants do not cross the phyto-geographical boundaries between Malesia and neighboring continents. At the Thai-Malaysian (Isthmus of Kra) border, for instance, 200 genera of the Asiatic flora do not cross further south, and, conversely, 375 genera of the Malesian flora do not occur north of the border, giving a so-called demarcation knot of 575 genera (van Steenis 1950). The boundary between Papua New Guinea and Australia is even stronger; the demarcation knot is 984 genera. Intermediate is the boundary between the Philippines and the Sino-Japanese flora (the demarcation knot is 686 genera). Although the endemicity at the generic and familial level is relatively low, the Malesian flora has a great number of endemic species, usually confined to one or a group of islands in the region (table 2.3). Ng, Low, and Mat Asri (1990) have shown that in Peninsular Malaysia there are about 2,830 species of trees, of which 746, or 26%, are endemic. Of these endemics, Ng, Low, and Mat Asri considered 301, or 40%, hyper-endemic species, recorded in only one of the states in the peninsula. As Soepadmo (1991, 1992) has shown, most of these, and other endemic species, have been recorded only from localities outside the strictly protected forest area of Taman Negara, Pahang. Severe

Table 2.3   Level of species endemism of selected Malesian families

Family	Genera/species in the world (N)	Genera in Malesia/ endemic (N)	Species in Malesia/ endemic (N)	Species endemic to Malesia (%)
Anacardiaceae	70 /600	21 / 2	150 / 110	73
Araucariaceae	2 / 40	2 / 0	13 / 11	85
Balanophoraceae	18 /45	4 / 1	10 / 6	60
Bombacaceae[a]	5 /45	4 / 3	44 / 43	95
Burseraceae	16 /550	8 / 1	107 / 100	93
Dipterocarpaceae	16 /507	10 / 3	386 / 346	90
Ericaceae	125 /3,500	12 / 4	740 / 722	98
Fagaceae	7 /700	5 / 0	180 / 162	90
Magnoliaceae	7 / 200	5 / 1	33 / 27	82
Nepenthaceae	1 / 71	1 / 0	63 / 62	98
Podocarpaceae	13 / 172	2 / 0	13 / 11	85
Ulmaceae	15 / 200	6 / 0	27 / 11	41

Sources: Data compiled from Ashton 1982; de Laubenfels 1988; Hansen 1976; Hou 1978; Leenhouts 1956; Sleumer 1966, 1967; Soepadmo 1972, 1977a; and Kurata 1976.
[a] Including only the tribe Durionae.

destruction or complete elimination of their natural habitats elsewhere will thus inevitably lead to their extinction. On a regional and global basis, continuous and extensive elimination of the species-rich lowland and hill dipterocarp forests of Malesia will also cause massive extinction of species of at least 962 genera with centers of distribution and speciation in the region. Examples are dipterocarp timber genera such as *Anisoptera, Cotylelobium, Dipterocarpus, Dryobalanops, Neobalanocarpus, Parashorea, Shorea, Upuna,* and *Vatica,* as well as such economically important and biologically interesting genera as *Artocarpus, Calamus, Durio, Lansium, Mangifera, Nepenthes,* and *Rafflesia.* As a center of species distribution, the tropical rainforests of Malesia play a vital role in the preservation of many economically valuable plant genetic resources; tropical timbers, rattans, and medicinal plants are a few examples.

THE ECONOMIC IMPORTANCE OF SOUTHEAST ASIAN SPECIES

Recent surveys conducted by botanists involved in the PROSEA (Plant Resources of Southeast Asia) project indicate that no fewer than 5,952 of the 40,000 species of vascular plants known to occur in Malesia have been reported to have actual or potential economic value (table 2.4).

Table 2.4 Selective economic uses of Southeast Asian tropical rainforest plants

Products / Commodity groups	Species (N)
Timber trees	1,462
Medicinal plants	1,135
Ornamental plants	520
Edible fruits and nuts	389
Fibers	227
Rattans	170
Poisonous and insecticidal plants	147
Spices and condiments	110
Others	1,792
Total	**5,952**

Source: Jansen et al. 1991.

Timber Trees

In Peninsular Malaysia there are 2,830 tree species in 532 genera and 100 families (Ng, Low, and Mat Asri 1990), whereas in Sabah and Sarawak at least 3,500 species in 500 genera and 105 families have been reported (Anderson 1980; Burgess 1966). In Kalimantan (Indonesian Borneo) Whitmore, Tantra, and Sutisna (1989) have recorded over 2,692 species in 378 genera and 84 families. Of the 2,830 tree species in Peninsular Malaysia, 677, or about 20%, reach timber size (40 cm dbh or larger), and 402 of these have been listed as commercial species and graded into 53 timber groups and 4 main timber classes. The balance of 275 species constitute the so-called lesser-known (acceptable) timbers (Kochummen 1973). In Sabah, Burgess (1966) listed 400 tree species in 233 genera and 47 families as timber species; of these, about 150 species are currently being exploited commercially (*Forestry* 1989).

Given the number of tree species (1,462) in table 2.4, there seems to be ample room to diversify the use of timber species available in Malaysian forests. The tropical rainforest of Malesia could be considered as a storehouse or gene bank of many tropical timber species useful for future development in timber-related industries, especially those species that are not fully used at present and regarded as lesser-known timbers, but which might be important in the future. Such tree resources may constitute at least 40% of the forest stands. Because tropical timbers are undoubtedly the most important forest products of

the region and, if managed on a sustained-yield basis, could be indefinitely renewable, their adequate conservation and appropriate management are vital.

Medicinal Plants

Perry and Metzger (1980) have compiled a list of about 6,000 species
of vascular plants from Southeast Asia that have been reported to have thera-
peutic properties and have been widely used in various systems of medicine
practiced in the region. Among plant families of the Malesian rainforest commonly used in the preparation and prescription of traditional medicines are the
Annonaceae, Apocynaceae, Araceae, Dioscoreaceae, Euphorbiaceae, Lauraceae, Menispermaceae, Rubiaceae, Rutaceae, Simaroubaceae, and Zingiberaceae. A report by Farnsworth et al. (1985) estimates that 88% of people in
developing countries of Africa, Asia, and Latin America rely chiefly on traditional medicine for their health needs. As for modern medicine, Soejarto and
Farnsworth (1989) have stated that 121 prescription drugs worldwide are derived
from 95 species of higher plants. Of these, 39 species producing 47 drugs are
tropical plants. Because there are a number of modern diseases or symptoms
(herpes, AIDS, cancer, arthritis, muscular dystrophy, Parkinsonism, drug addic-
tion, cystic fibrosis, hemophilia, hypertension, and many others) for which satisfactory cures remain to be discovered and developed, and because only a
handful of the thousands of tropical plant species have been subjected to a detailed investigation for medicinal properties, tropical rainforest plants are potentially extremely valuable as a source of useful drugs and other compounds.

Ornamental Plants

Of the estimated 500 species of trees, shrubs, and herbs cultivated in
Malesia as ornamental plants, about 100 originated in the region. The others
were introduced from other tropical and subtropical countries, and some were
even brought from colder temperate regions. This is indeed surprising, because
surveys conducted by Soepadmo (1976a, 1976b, 1977b, 1977c); Phillipps,
Phillipps, and Phillipps (1982); and Weber and Kiew (1983) indicate that hundreds of plant species growing in the tropical rainforests of Malesia have potential as ornamentals. Many trees (species of *Calophyllum, Cinnamomum,
Gardenia, Polyalthia, Sterculia,* and several others) have graceful forms or produce attractive flowers or young flushes of leaves. Likewise, many herbaceous
plants either produce colorful flowers or foliage or have a unique growth form.
The orchid family alone has more than 6,500 species in Southeast Asia. Native
to the region, these species are well adapted to local soil and climate and are potentially economically valuable resources for the horticultural industry.

According to Lee (1991), the annual value of the world trade in flowers and other ornamental plant products is about $1.6 billion. This figure excludes domestic demands, which in Malaysia are worth about $6.3 million annually. It would thus be economically worthwhile to develop the ornamental plant resources of Southeast Asian forests.

Fruit Trees

Reports by Sastrapradja (1975) and Soepadmo (1979a) suggest that at least 124 fruit-tree species are cultivated in Malesia. Of these, only 23 species, or about 27%, are indigenous; the others originated from other tropical and subtropical countries, particularly in Central and South America. In contrast, investigations by Kostermans (1958, 1991), Saw et al. (1991), and others indicate that no fewer than 120 species of wild trees in the tropical forests of Malesia produce edible fruits. Notable examples are species in the genera *Artocarpus* (Moraceae), *Bouea*, *Mangifera* (Anacardiaceae), *Durio* (Bombacaceae), *Garcinia* (Clusiaceae), *Lansium* (Meliaceae), and *Nephelium*, *Pometia*, and *Xerospermum* (Sapindaceae). Often many fruit trees in the same genus occur together. At Pasoh Forest Reserve, for example, there are 13 species of *Garcinia* and 12 species of *Mangifera* (Saw et al. 1991). Like many other trees in the tropical rainforests, the wild fruit trees are often found at a very low density. These wild fruit-tree species are potentially valuable as genetic resources for future use, considering that 24 of them have been cultivated in orchards and villages, 38 are congeneric with cultivated fruit crops, and the others possess desirable genetic traits for breeding, selection, and improvement programs (Primack 1985). Therefore, if a local fruit industry is to be developed and expanded in the near future, the protection, conservation, and cultivation of wild fruit-tree species are extremely important. The Malaysian fruit industry is rapidly expanding, and in 1987 alone the area under fruit cultivation was estimated at 186,000 ha (*Malaysian Fruit Industry* 1989); however, Malaysia is still a net importer of fruits. In 1988, for example, the country imported temperate and tropical fruits valued at $68 million, and in the same period exported fruits worth only $39 million.

Rattans

These unique climbing, calamoid palms comprise about 600 species in 12 genera, distributed in tropical and subtropical Africa, India, south China, Burma, Thailand, Indochina, and Malesia. The center of distribution and species diversity of rattans is located in the Malesian tropical rainforests, where about 540 species in 10 genera have been reported. In Malaysia, rattans are represented by at least 200 species in 10 genera (Dransfield 1979, 1984, 1992), whereas in Indonesia more than 300 species in 7 genera have been recorded

(Mogea 1991; Silitonga 1985). Because of consumer preferences and availability of large quantities of particular species, however, no more than 20 species have been harvested and sold commercially. The most important of these are *Calamus caesius, C. inops, C. manan, C. scipionium,* and *C. trachycoleus.* In 1985, Indonesia alone exported 137,000–155,000 tons of unprocessed canes valued at $97 million, and in the same period the world trade value for rattan furniture and other products was estimated to be $2.7 billion. The rattan industries' socioeconomic contribution to the producing countries is indeed significant. Because of overexploitation of preferred species (such as *Calamus inops* and *C. manan*) and loss of natural habitat for the others, however, a number of species may become extinct before their economic potential for cultivation is realized (Kiew 1991a; Mogea 1991; Pearce 1991). The rattan-based industries in producing countries of Malesia are expanding rapidly, and the foreign exchange earnings generated by such activities are increasing every year. In 1987, Indonesia, the Philippines, Thailand, and Malaysia exported rattan furniture and other finished products valued at $215 million, $85 million, $23 million, and $11 million, respectively. Because only 20,000–25,000 ha of rattan plantations have been established in the region so far, efforts should be made to protect and conserve representatives of rattan species and their natural habitats so that their future contribution toward the socioeconomic development of the producing countries can be sustained. Genetic improvements and better cultivation techniques will certainly result in greater yields of rattans from plantations.

The selected examples of forest products clearly demonstrate that the tropical rainforests of Malesia contain diverse renewable natural resources other than tropical timber. Although the economic value of timber and related products is undeniable, nontimber products are also socioeconomically significant. Surveys made by de Beer and McDermott (1989) show that the livelihood of at least 29 million people in Malesia is critically dependent on these nontimber forest products, and the total population benefiting from these resources is even greater. The surveys also indicate that forest resources account for most of the several billion dollars in annual world trade of nontimber products from Malesia, including nearly $3 billion in finished rattan products alone.

CONSERVATION

Efforts to protect and conserve adequate representatives of tropical rainforest biodiversity in Malesia have met with mixed success (Kiew 1990; Leong et al. 1992; Soepadmo 1979b, 1987). Many social and scientific factors contribute to this problem: the increasing rate of exploitation and the lack of political will and commitment combined with an incomplete knowledge of the ecological func-

Table 2.5 Protected forest habitats in selected countries of Malesia

Country	Land area (ha)	Remaining forest habitat in ha (% of total land area)	IUCN category	Existing protected forest habitat in ha (% of total remaining forest)
Brunei	576,500	471,500 (82)	IV	122,367 (23)
Indonesia	191,944,300	104,002,600 (54)	I-V	17,799,787 (17)
Malaysia	32,860,000	19,370,000 (59)	I-V	1,710,931 (9)
Papua New Guinea	46,284,000	39,048,400 (84)	II & IV	29,016 (.07)
Philippines	30,000,000	6,971,200 (23)	II-V	583,999 (8)
Total	**301,664,800**	**169,863,700[a] (56)**		**20,246,100 (12)**

Sources: Data from *Forestry* 1984, 1989, 1991; IUCN 1990; Mohd. Khan 1991; Nais and Ali 1991; Ngui 1991; and Othman 1991.

[a] Recent estimate given by Collins, Sayer, and Whitmore (1991) and Whitmore and Sayer (1992) = 180,017,000 ha.

tioning (carrying capacity) of the forest, the population dynamics and reproductive biology of its component species, and the mechanisms that maintain the high species diversity through time and space. The area of remaining natural forests designated as protected habitats in Malesian countries is grossly inadequate (table 2.5). Of the estimated 170 million ha of remaining natural forest habitats in Brunei, Indonesia, Malaysia, the Philippines, and Papua New Guinea, only about 20.25 million ha, or approximately 12%, are officially designated national parks or other protected categories. These protected areas constitute only about 6.7% of the original land area and in most cases do not include the whole range of vegetation types originally present in the region.

Most species of tropical forest organisms are present at very low densities and are randomly distributed throughout their geographical range. Many species, particularly large vertebrates, do not thrive in disturbed or degraded habitats; a relatively large, contiguous area of undisturbed forest is required to maintain a viable breeding population of such species. The relatively small area of protected, undisturbed forest cannot capture the entire range of biodiversity (from genetic to organismic and ecosystem level) of the region. In the case of Malaysia, for instance, the total area of protected forests consists of 793,427 ha of national and state parks, 107,304 ha of virgin jungle reserves, 628,534 ha of wildlife reserves and bird sanctuaries, 99,911 ha of protection forests, and 81,755 ha of other conservation areas, for a total of 1,710,931 ha of protected forests (table 2.6). Natural vegetation on limestone and ultramafic rock formations,

Table 2.6 Existing protected forest habitats in Malaysia

Type of protected habitat	Peninsular Malaysia (ha)	Sabah (ha)	Sarawak (ha)	Total (ha)
National/State parks	434,300	245,172	113,955	793,427
Virgin jungle reserves	19,000	88,304	—	107,304
Wildlife preserves/ Bird sanctuaries	310,000	143,682	174,851	628,543
Protection forests	—[a]	99,911	—[a]	99,911
Other conservation areas	—	81,755	—	81,755
Total protected areas	**763,300**	**658,824**	**288,806**	**1,710,931**
Total land area	**13,160,000**	**7,400,000**	**12,300,000**	**32,860,000**
Total remaining forest	**6,150,000**	**4,520,000**	**8,700,000**	**19,370,000**
Total PFE	**4,750,000**	**3,350,000**	**6,000,000**	**14,100,000**

Sources: Data compiled from *Forestry* 1989, 1991; Mohd. Khan 1991; Nais and Ali 1991; Ngui 1991; and Othman 1991. See also Abang Morshidi and Gumal, this volume.

[a] According to Othman (1991) about 1.9 million ha of forest in Peninsular Malaysia and 1.4 million ha in Sarawak (only 500,000 ha, according to Abang Morshidi and Gumal, this volume) are designated protected forest. Because its legal and conservation status cannot be ascertained at the moment, however, the area is not included here.

[b] Protected areas are 5.2% of total land area, 8.8% of total remaining forest, and 12.1% of total forest under permanent forest estate.

[c] Rounded figures.

heath, mangrove, freshwater, and peatswamp forests are all absent from these protected areas or inadequately represented.

In Peninsular Malaysia, of the roughly 763,301 ha of protected forest habitats, only the Taman Negara (approximately 434,300 ha, of which only about 30% is lowland dipterocarp forest) can be considered conservationally secure. The virgin jungle reserves (roughly 19,000 ha) consist of forest fragments, each measuring between 3 and 1,600 ha. Surveys by Putz (1978) suggest that most of these jungle reserves have been reduced in size or transferred to other land uses, making them ineffective as conservation areas. Likewise, in the so-called wildlife or bird sanctuaries totaling about 310,000 ha, the forests per se are not usually strictly protected, although a number of animal species living within the area are off-limits for hunting and collecting. Efforts by several government departments and nongovernment organizations (NGOs) to increase the area of totally protected forest (to include the Rompin-Endau and Belum forests, for example) have met with reluctance from the concerned authorities.

This reluctance is unfortunate, because the Taman Negara represents only forest habitats of the central region of the peninsula, and studies by Ng, Low, and Mat Asri (1990) on endemic tree species and by Dransfield and Kiew (1990) on palm species have shown that this particular national park harbors only about 3% of the endemic tree species and about 30% of the palm species known in Peninsular Malaysia (table 2.7).

In Sabah the situation is very similar. The three major parks (Crocker Range, Mt. Kinabalu, and Tawau Hill) cover a total area of approximately 243,261 ha, consisting mainly of hill and montane forests. Extremely species-rich forests on ultramafic outcrops are not adequately protected (Beaman and Beaman 1990). The virgin jungle reserves, amounting to about 88,304 ha, consist of 39 separate forest patches ranging in size from 24 to 18,000 ha. How well these reserves are protected and managed is largely unknown. There are only two sizable wildlife reserves in the state of Sabah, namely, the Kelumba Wildlife Reserve (20,383 ha) and the Tabin Wildlife Reserve (123,000 ha). The former is mainly composed of swamp forest, and the latter consists primarily of lowland and hill dipterocarp forests. In Sabah, as in Peninsular Malaysia, the emphasis is on the protection of wild animal species rather than on the forest ecosystem as a whole. Within the Sabah Foundation's forest concession lands are two conservation areas, the Danum Valley Area (42,755 ha) and the Mt. Lotung Area (39,000 ha). Their legal status as permanent conservation areas is not clearly established. Also, the protection forest reserve, defined as "forest conserved for the maintenance of the stability of essential climatic, watershed, and other environmental factors" (*Forestry* 1989), covers an area of about 100,000 ha, consisting of 26 isolated forest sites that range in size from 11 to 22,700 ha. Although timber harvesting is prohibited in such forests, shifting cultivators and illegal loggers often encroach.

The status of forest habitat conservation in the state of Sarawak is not much better. According to Ngui (1991), until 1989 strictly protected natural forests in Sarawak totaled about 253,940 ha, representing about 2% of the total land area and 3% of the remaining natural forest cover in the state. These forests consist of seven national parks, each ranging in size from 2,230 to 52,866 ha, for a total of 79,089 ha, and three wildlife sanctuaries, ranging in size from 1.4 to 168,758 ha and totaling 174,851 ha. By 1991, two additional national parks, Batang Ai and Loagan Bunut, with a total area of 34,776 ha, were designated (*Forestry* 1991). As in the other two regions, the protected areas are composed mainly of lowland and hill dipterocarp forests. Among the inadequately protected forest habitats in the state are the mangroves, freshwater and peatswamp forests, upper montane forests, and specialized vegetation on ultramafic and other

Table 2.7   Species in Taman Negara National Park and in all Peninsular Malaysia

	Taman Negara (area = 434,300 ha) (*N*)	Peninsular Malaysia (land area = 13,160,000 ha) (*N*)
Fruit trees	43	100
Palms	73	243
Endemic trees	22	746
Limestone flora	125	1,216
Ferns	134	500
Small mammals	71	194
Birds	260	500
Freshwater fish	109	250
Amphibians	55	89
Snakes	67	116
Butterflies	248	1,034

Source: Data compiled from various chapters in Kiew 1991b.

unique rock and soil formations (Bruenig 1991; Ngui 1991). The state authorities, however, are making serious efforts to enlarge the acreage of totally protected areas to almost 1 million ha, representing about 8% of the total land area or about 11.5% of the natural forests remaining in Sarawak (*Forestry* 1991; Abang Morshidi and Gumal, this volume; Ngui 1991).

DISCUSSION AND CONCLUSIONS

Malesian tropical rainforests are centers of distribution and species diversity, not only for tropical hardwood timber species, but also for many other plant groups with economic value, such as fruit trees, ornamental plants, rattans, and medicinal plants. In phytogeographical terms, because of their unique geological history and geographical position, the forests of the region are one of the world's centers of origin, diversity, and endemism of many scientifically interesting plant genera, including the dipterocarps, *Calamus, Nepenthes, Trigonobalanus* (Fagaceae), *Rafflesia,* and many others. Van Steenis (1987) has stated that 2,382 genera of seed-bearing plants are known in Malesia, of which about 40% are either endemic or have their center of distribution and species diversity in the region. The degree of endemism is even higher at the species level.

Ecologically, the tropical rainforests are vital in regulating local, regional, and global climate (Goreau and de Mello 1988; Keller et al. 1989; Prance

t and Matson 1992). These forests, particularly the lowland and
p forests, have among the highest standing biomass per unit area
1 the world, with a range of 200–600 t ha^{-1} (Soepadmo 1993).
the organic carbon stored in the soils, this plant biomass consti-
...... one of the most important sources of carbon in the world's vegetation
(Houghton, this volume). Further, at both local and regional levels the forest ex-
erts a great influence on the stability and maintenance of hydrological and heat
balance. With four or five canopy layers, the forest has a high rate of rainfall in-
terception (20–40%), thus reducing the amount and direct impact of rainwater
on the topsoil. The porous soils under the forest canopy facilitate infiltration
and percolation of water into the soils, reducing surface runoff and soil erosion.
Under a forest, excess rainwater will flow gently through the soil layers and
eventually discharge into the streams and rivers as clear water. Without this forest cover, rainwater runs rapidly off the land, causing soil erosion and resulting
water pollution, as well as flooding. A study by Aoki, Yabuki, and Koyama (1975)
indicates that air temperature of a forested area is about 3–5° C lower than that
of a surrounding nonforested area. During a hot day, this difference in air temperature generates air movement from the atmosphere above the forest to the
surrounding warmer area, thus ameliorating the heat-island effect of the cleared
area. The process reverses at night: the wind blows from the cooler settled area
to the warmer forested area, dispersing pollutants suspended in the urban atmosphere. On a regional and global basis, the presence of sizable dense forests
with a high standing biomass regulates the release of CO_2 from the biosphere to
the atmosphere, thus reducing the impact of the so-called greenhouse effect and
global warming.

The strictly protected forest areas in the Malesian countries are
grossly inadequate to capture and preserve the range of biodiversity in the region, except perhaps for certain groups of vertebrate and bird populations
(A. D. Johns 1983, 1986a, 1986b, 1988; MacKinnon and MacKinnon 1986;
Wilson and Johns 1982). Connor and McCoy's (1979) equation of species-area
relations can be used to predict the number of species that eventually will become extinct if the rate of deforestation is maintained. According to this equation, the area of strictly protected forest in Malesia (20.25 million ha) will at
most capture and preserve about 50% of the original species diversity. This figure will be even lower if we consider that, in most countries, the protected areas
consist of fragmented forest patches and normally do not include the full range
of natural habitats available in the regional landscape (Terborgh 1992). Rare and
hyper-endemic species confined to specialized habitats (limestone hills, ultramafic outcrops, and so on) are particularly vulnerable to extinction. Once such

specialized habitats, many of which are presently excluded from conservation areas, are destroyed, the entire population of such species will be lost forever. Because the available data clearly indicate that most species indigenous to the tropical rainforests of Malesia belong to this category, massive species extinctions are a very real possibility. It is imperative that we, the nations of the region, should seriously endeavor to manage and conserve adequate representatives of the varied biological resources available in our native rainforests, for the benefit of present and future generations. In view of this problem, and in order to enable countries in the Malesian region to fulfill the commitments outlined in the 1992 Rio Convention on Biodiversity, the following steps are recommended:

1. Develop national strategies, plans, or programs for the conservation and sustainable use of biological diversity in the country.
2. For in situ conservation, increase the area of strictly protected forest habitats to at least 20% of the remaining natural forest area, and increase production forest area to at least 50% of the total land area.
3. Reduce the rate of timber harvesting to a sustainable level as determined by the National Forestry Policy.
4. Incorporate biodiversity conservation programs into the management of the production forests.
5. Establish national centers and facilities for ex situ conservation of threatened and endangered plant and animal species. Such facilities include botanic gardens, public parks, arboretums, seed banks, zoological parks, and wildlife rehabilitation centers.
6. Intensify ecological research, surveys, and inventories both inside and outside the boundaries of the existing national parks, wildlife sanctuaries, and other protected areas to obtain comprehensive data pertaining to the diversity, distribution, ecological amplitude, conservation status, and habitat preferences of the component species, as well as the responses of these species to human and natural disturbances.
7. Carry out rescue operations at those forested areas that have been designated for conversion to nonforestry land uses. These operations aim to collect planting materials for species to be preserved, cultivated, and propagated in ex situ conservation centers, and to capture and transfer threatened animal species to rehabilitation centers to be protected and raised.
8. Establish a national institute of biodiversity to intensify and coordinate research activities pertaining to the conservation, sustainable use, and management of biological resources of the country.

9. Enhance public awareness of the importance and urgency of protecting and conserving biodiversity in the country.

REFERENCES

Abang Morshidi, A. H. K., and M. T. Gumal. This volume. The role of totally protected areas in preserving biological diversity in Sarawak.

Anderson, J. A. R. 1963. The flora of the peatswamp forests of Sarawak and Brunei including a catalogue of all recorded species of flowering plants, ferns and fern allies. *Gard. Bull. Singapore* 20: 131.

———. 1980. *A Checklist of the Trees of Sarawak.* Sarawak Forest Department, Kuching.

Aoki, M., K. Yabuki, and H. Koyama. 1975. Micrometeorology and assessment of primary production of a tropical rain forest in West Malaysia. *J. Agr. Met.* 31: 115.

Ashton, P. S. 1982. Dipterocarpaceae. *Flora Malesiana* Ser. I, 9: 237.

———. 1984. Biosystematics of tropical forest plants: A problem of rare species. In *Plant Biosystematics.* W. F. Grant, editor. Academic Press, New York.

Ashton, P. S., T. J. Givnish, and S. Appanah. 1988. Staggered flowering in the Dipterocarpaceae: New insight into floral induction and the evolution of mast flowering in the aseasonal tropics. *Amer. Nat.* 132: 44.

Beaman, J. H., and R. S. Beaman. 1990. Diversity and distribution patterns in the flora of Mount Kinabalu. In *The Plant Diversity of Malesia.* P. Baas, K. Kalkman, and R. Geesing, editors. Kluwer Academic, Boston.

Beaman, R. S., J. H. Beaman, C. W. Marsh, and P. V. Woods. 1985. Drought and forest fires in Sabah in 1983. *Sabah Soc. J.* 8: 10.

Brown, S., and A. L. Lugo. 1982. The storage and production of organic matter in tropical forests and their role in the global carbon cycle. *Biotropica* 14: 161.

Bruenig, E. F. 1991. Kerangas and kerapah forests of Sarawak. In *The State of Nature Conservation in Malaysia.* R. Kiew, editor. Malayan Nature Society, Kuala Lumpur.

Burgess, P. F. 1966. *Timbers of Sabah.* Sabah Forest Record No. 6. Forest Department, Sabah.

Chin, S. C. 1977. The limestone flora of Malaya, I. *Gard. Bull. Singapore* 30: 165.

———. 1979. The limestone flora of Malaya, II. *Gard. Bull. Singapore* 32: 64.

———. 1983a. The limestone flora of Malaya, III. *Gard. Bull. Singapore* 35: 137.

———. 1983b. The limestone flora of Malaya, IV. *Gard. Bull. Singapore* 36: 31.

Collins, N. M., J. A. Sayer, and T. C. Whitmore, editors. 1991. *The Conservation Atlas of Tropical Forests: Asia and the Pacific.* Macmillan, London.

Connor, E. F., and E. D. McCoy. 1979. The statistics and biology of the species-area relationship. *Amer. Nat.* 113: 791.

Croat, T. 1978. *Flora of Barro Colorado Island.* Stanford University Press, Stanford.

de Beer, J. H., and M. J. McDermott. 1989. *The Economic Value of Non-timber Forest Products in Southeast Asia.* Netherlands Committee for IUCN, Amsterdam.

de Laubenfels, D. J. 1988. Coniferales. *Flora Malesiana* Ser. I, 10: 337.

Deteriler, R. P., and C. A. S. Hall. 1988. Tropical forests and the global carbon cycle. *Science* 239: 42.

Dransfield, J. 1979. *A Manual of the Rattans of the Malay Peninsula.* Malayan For. Rec. No. 29. Forest Department of Peninsular Malaysia, Kuala Lumpur.

———. 1984. *The Rattans of Sabah.* Sabah Forest Department, Sandakan.

———. 1992. *The Rattans of Sarawak.* Royal Botanic Gardens, Kew, Surrey, England, and Sarawak Forest Department, Kuching.

Dransfield, J., and R. Kiew. 1990. The palms of Taman Negara. *J. Wildlife & Parks* 10: 38.

Farnsworth, N. R., O. Okerele, A. S. Bingel, D. D. Soejarto, and Zhengang Guo. 1985. Medicinal plants in therapy. *Bull. WHO* 63: 965.

Flenley, J. R. 1979. *The Equatorial Rain Forests: A Geological History*. Butterworth, London.

―――. 1984. Late quaternary changes of vegetation and climate in the Malesian mountains. *Erdwiss Fors* 18: 262.

Forestry in Malaysia. 1984. Ministry of Primary Industries, Kuala Lumpur.

Forestry in Sabah. 1989. Sabah Forest Department, Sandakan.

Forestry in Sarawak. 1991. Sarawak Forest Department, Kuching.

Gentry, A. H., and C. Dodson. 1987. Contribution of non-tree species richness of a tropical rain forest. *Biotropica* 19: 149.

Goreau, T. J., and W. Z. de Mello. 1988. Tropical deforestation: Some effects on atmospheric chemistry. *Ambio* 17: 275.

Hansen, B. 1976. Balanophoraceae. *Flora Malesiana* I, 7: 783.

Hartshorn, G. 1983. Plants: Introduction. In *Costa Rican Natural History*. D. H. Janzen, editor. University of Chicago Press, Chicago.

Ho, C. C., D. McC. Newbery, and M. E. D. Poore. 1987. Forest composition and inferred dynamics in Jengka Forest Reserve, Malaysia. *J. Trop. Ecol.* 3: 25.

Hou, D. 1978. Anacardiaceae. *Flora Malesiana* Ser. I, 8: 395.

Houghton, R. A. This volume. Tropical forests and climate.

IUCN (International Union for the Conservation of Nature and Natural Resources). 1990. *1990 United Nations List of National Parks and Protected Areas*. IUCN, Gland, Switzerland.

Jansen, P. C. M., R. H. M. J. Lemmens, L. P. A. Oyen, J. S. Siemonsma, F. M. Satvest, and J. L. C. H. van Valkenburg, editors. 1991. *Plant Resources of Southeast Asia—Basic List of Species and Commodity Grouping*. Pudoc, Wageningen, The Netherlands.

Johns, A. D. 1983. Wildlife can live with logging. *New Scientist* 99: 206.

―――. 1986a. Effects of selective logging on the behaviour ecology of West Malaysian primates. *Ecology* 67: 684.

―――. 1986b. Effects of selective logging on the ecological organization of Peninsular Malaysian avifauna. *Forktail* 1: 65.

―――. 1988. Effects of "selective" timber extraction on the rain forest structure and composition and some consequences for frugivores and folivores. *Biotropica* 20: 31.

Johns, R. J. 1986. The instability of the tropical ecosystem in New Guinea. *Blumea* 34: 21.

―――. 1989. The influence of drought on tropical rain forest vegetation in Papua New Guinea. *Mt. Res. & Devel.* 9: 248.

―――. 1990. The illusionary concept of climax. In *The Plant Diversity of Malesia*. P. Baas, K. Kalkman, and R. Geesing, editors. Kluwer Academic, Boston.

Keller, M., D. J. Jacob, S. C. Wofsy, and R. C. Harriss. 1989. Effects of tropical deforestation on global and regional atmospheric chemistry. *Climatic Change* 19: 139.

Kiew, R. 1990. Conservation of plants in Malaysia. In *The Plant Diversity of Malesia*. P. Baas, K. Kalkman, and R. Geesing, editors. Kluwer Academic, Boston.

―――. 1991a. Palm utilization and conservation in Peninsular Malaysia. In *Palms for Human Needs in Asia*. D. Johnson, editor. WWF/IUCN Project No. 3325. A. A. Balkema, Rotterdam.

Kiew, R., editor. 1991b. *The State of Nature Conservation in Malaysia*. Malaysian Nature Society, Kuala Lumpur.

Kochummen, K. M. 1973. Lesser known timber trees of Malaysia. In *Proc. Symp. Biol. Res. & Nat. Dev.* E. Soepadmo and K. G. Singh, editors. Malaysian Nature Society, Kuala Lumpur.

Kochummen, K. M., J. V. LaFrankie, Jr., and N. Manokaran. 1990. Floristic composition of Pasoh Forest Reserve, a lowland rain forest in Peninsular Malaysia. *J. Trop. For. Sci.* 3: 1.

Kostermans, A. J. G. H. 1958. The genus *Durio* Adans. (Bombacaceae). *Reinwardtia* 4: 357.

―――. 1991. *Kedondong, Ambarella, Amra: The Spondiadeae (Anacardiaceae) in Asia and the Pacific Area*. Bina Karya 78 Printing Work, Bogor, Indonesia.

Kurata, S. 1976. *Nepenthes of Mount Kinabalu*. Sabah National Parks, Sabah.

Lee, S. K. 1991. Development of floriculture industry: Challenges and prospects. *Malay. Orchid Bull.* 5: 57.

Leenhouts, P. W. 1956. Burseraceae. *Flora Malesiana* Ser. I, 5: 209.

Leong, Y. K., G. Davison, B. H. Kiew, M. Koshokumar, S. W. Lee, and M. T. Lim. 1992. An approach to conservation in Malaysia: The Malaysian Nature Society second blueprint for conservation. *Malay. Nat. J.* 45: 438.

Mabberley, D. J. 1987. *The Plant Book: A Portable Dictionary of the Higher Plants.* Cambridge University Press, Cambridge.

MacKinnon, J., and K. MacKinnon. 1986. *Review of the Protected Area Systems in the Indo-Malayan Realm.* IUCN, Gland, Switzerland.

Malaysian Fruit Industry: Directory 1989/90. 1989. Staff Coop. Ministry of Agriculture, Kuala Lumpur.

Mogea, J. P. 1991. Indonesia: Palm utilization and conservation. In *Palms for Human Needs in Asia.* D. Johnson, editor. WWF/IUCN Project No. 3325. A. A. Balkema, Rotterdam.

Mohd. Khan, M. K. 1991. The role of protected areas in the conservation of biological diversity. *Malay. Nat. J.* 45: 238.

Morley, R. J., and J. R. Flenley. 1987. Late cainozoic vegetational and environmental changes in the Malay Archipelago. In *Biogeographical Evolution of the Malay Archipelago.* T. C. Whitmore, editor. Clarendon Press, Oxford.

Nais, J., and L. Ali. 1991. Sabah parks. In *The State of Nature Conservation in Malaysia.* R. Kiew, editor. Malaysian Nature Society, Kuala Lumpur.

Newbery, D. McC., and J. Proctor. 1984. Ecological studies in four contrasting lowland rain forests in Gunung Mulu National Park, Sarawak, IV: Association between tree distribution and soil factors. *J. Ecol.* 72: 475.

Newsome, J., and J. R. Flenley. 1988. Late quaternary vegetational history of the Central Highlands of Sumatra, II: Palaeo-palinology and vegetational history. *J. Biogeogr.* 15: 555.

Ng, F. S. P., C. M. Low, and Mat Asri N. S. 1990. *Endemic Trees of the Malay Peninsula.* Res. Pamph. No. 106. FRIM, Kuala Lumpur.

Ngui, S. K. 1991. National parks and wildlife sanctuaries in Sarawak. In *The State of Nature Conservation in Malaysia.* R. Kiew, editor. Malaysian Nature Society, Kuala Lumpur.

Othman, A. M. 1991. Forest resources in Malaysia. In *Environment and Development: Malaysian Perspectives.* S. Thanarajasingam et al., editors. Ministry of Foreign Affairs, Kuala Lumpur.

Palm, C. A., R. A. Houghton, and J. M. Melillo. 1986. Atmospheric carbon dioxide from deforestation in Southeast Asia. *Biotropica* 18: 177.

Pearce, K. G. 1991. Palm utilization and conservation in Sarawak (Malaysia). In *Palms for Human Needs in Asia.* D. Johnson, editor. WWF/IUCN Project No. 3325. A. A. Balkema, Rotterdam.

Perry, L. M., and J. Metzger. 1980. *Medicinal Plants of East and Southeast Asia: Attributed Properties and Uses.* MIT Press, Cambridge.

Phillipps, A., S. M. Phillipps, and C. Phillipps. 1982. Some ornamental plants of Sabah. *Nat. Malaysiana* 7: 20.

Poore, M. E. D. 1968. Studies in Malaysian rain forest, I: The forest on triassic sediments in Jengka forest reserve. *J. Ecol.* 56: 143.

Prance, G. T., editor. 1986. *Tropical Rain Forests and the World Atmosphere.* Westview, Boulder, Colo.

Primack, R. 1985. Comparative studies of fruits in wild and cultivated trees of chempedak (*Artocarpus integer*) and terap (*A. odoratissimus*) in Sarawak, East Malaysia, with additional information on the reproductive biology of the Moraceae in Southeast Asia. *Malay. Nat. J.* 39: 1.

Primack, R., and P. Hall. 1991. Species diversity research in Bornean forests with implications for conservation biology and silviculture. *Tropics* 1: 91.

Putz, F. E. 1978. *A Survey of Virgin Jungle Reserves in Peninsular Malaysia*. Res. Pamph. No. 73. FRIM, Kuala Lumpur.

Roos, M. C. 1992. Flora Malesiana action plan. *Second Flora Malesiana Symposium: Programme and Summary of Papers and Posters*. Yogyakarta, Indonesia.

Sastrapradja, S. 1975. Tropical fruit germplasms in Southeast Asia. In *Southeast Asian Plant Genetic Resources*. J. T. Williams, C. H. Lamoureux, and N. Wulijarni-Soetjipto, editors. IBPGR/SEAMEO-BIOTROP/LIPI, Bogor, Indonesia.

Saw, L. G., J. V. LaFrankie, K. M. Kochummen, and S. K. Yap. 1991. Fruit trees in a Malaysian rain forest. *Econ. Bot.* 45: 120.

Silitonga, T. 1985. Country report: Indonesia. In *Proc. Rattans Seminar*. K. M. Wong and N. Manokaran, editors. Rattan Information Centre, FRIM, Kuala Lumpur.

Sleumer, H. 1966. Ericaceae. *Flora Malesiana* Ser. I, 6: 469.

———. 1967. Ericaceae. *Flora Malesiana* Ser. I, 6: 669.

Soejarto, D. D., and N. R. Farnsworth. 1989. Tropical rain forest: Potential source of new drugs? *Persp. Biol. & Med.* 32: 244.

Soepadmo, E. 1972. Fagaceae. *Flora Malesiana* Ser. I, 7: 265.

———. 1976a. Wild orchids. *Nat. Malaysiana* 1: 26.

———. 1976b. Ginger plants. *Nat. Malaysiana* 1: 32.

———. 1977a. Ulmaceae. *Flora Malesiana* Ser. I, 8: 31.

———. 1977b. Pitcher plants. *Nat. Malaysiana* 2: 38.

———. 1977c. The Aroids. *Nat. Malaysiana* 2: 12.

———. 1979a. Genetic resources of Malaysian fruit trees. *Malaysian Appl. Biol.* 8: 33.

———. 1979b. The role of tropical botanical gardens in the conservation of threatened valuable plant genetic resources in Southeast Asia. In *Survival or Extinction*. H. Synge and H. Townsend, editors. Royal Botanic Gardens, Kew, Surrey, England.

———. 1983. *Forest and Man: An Ecological Appraisal*. Inaugural Lecture. University of Malaya, Kuala Lumpur.

———. 1987. The impacts of man's activities on the unique floras of Malaysian mountains. In *Proc. Reg. Workshop on Impact of Man's Activities on Tropical Upland Forest Ecosystems*. H. Yusuf, A. Kamis, Nik Muhamad Majid, and M. Shukri, editors. Faculty of Forestry, UPM, Serdang, Malaysia.

———. 1991. Indomalayan wild orchids: Their ecological and conservation status. *Malay. Orchid Bull.* 5: 63.

———. 1992. Conservation status of medicinal plants in Peninsular Malaysia. In *Medicinal Products from Tropical Rain Forests*. S. Khozirah, A. K. Azizol, and M. A. Abd Razak, editors. FRIM, Kuala Lumpur.

———. 1993. Tropical rain forest as carbon sink. *Chemosphere* 27: 1025.

Terborgh, J. 1992. Maintenance of diversity in tropical forests. *Biotropica* 24: 283.

van Steenis, C. G. G. J. 1950. The delimitation of Malaysia and its plant geographical division. *Flora Malesiana* Ser. I, 1: lxxib–lxxiia.

———. 1969. Plant speciation in Malesia, with special reference to the theory of non-adaptive saltatory evolution. *Biol. J. Linn. Soc.* 1: 97.

———. 1979. Plant geography of East Malesia. *Biol. J. Linn. Soc.* 79: 97.

———. 1987. *Checklist of Generic Names in Malesian Botany (Spermatophytes)*. Flora Malesiana Foundation, Leiden, The Netherlands.

van Welzen, P. C. 1992. Species richness and speciation in Malesia. *Second Flora Malesiana Symposium: Programme and Summary of Papers & Posters*. Yogyakarta, Indonesia.

Vitousek, P. M., and P. A. Matson. 1992. Tropical forests and trace gases: Potential interaction between tropical biology and atmospheric sciences. *Biotropica* 24: 233.

Weber, A., and R. Kiew. 1983. Gesneriads of Peninsular Malaysia. *Nat. Malaysiana* 8: 24.

Whitmore, T. C., editor. 1981. *Wallace's Line and Plate Tectonics*. Clarendon Press, Oxford.

————. 1987. *Biogeographical Evolution of the Malay Archipelago*. Clarendon Press, Oxford.

Whitmore, T. C., and J. A. Sayer. 1992. Deforestation and species extinction in tropical moist forests. In *Tropical Deforestation and Extinction*. T. C. Whitmore and J. A. Sayer, editors. Chapman and Hall, London.

Whitmore, T. C., I. G. M. Tantra, and U. Sutisna. 1989. *Tree Flora of Indonesia: Checklist for Kalimantan*. Parts I and II. FRDC, Bogor, Indonesia.

Wilson, W. L., and A. D. Johns. 1982. Diversity and abundance of selected animal species in undisturbed forest, selectively logged forest and plantations in East Kalimantan, Indonesia. *Biol. Cons.* 24: 205.

Yamakura, T., A. Hagihara, S. Sukardjo, and H. Ogawa. 1986. Tree size in a mature dipterocarp forest stand in Sebulu, East Kalimantan, Indonesia. *Southeast Asian Stud.* 23: 452.

3   STRUCTURE, DYNAMICS, AND MANAGEMENT OF
RAINFORESTS ON NUTRIENT-DEFICIENT SOILS
IN SARAWAK

Eberhard F. Bruenig and Hans J. Droste

Eberhard F. Bruenig and Hans J. Droste

SITE AND SOIL

The fragility of certain nutrient-deficient tropical soils, such as podzols, and the constraints this fragility imposes on land use are well documented (Bruenig and Klinge 1969; Klinge 1968). Land misuse and abuse by agriculture and forestry, however, continue unabated worldwide. In Sarawak these fragile soils are primarily stagnant clay soils, excessively draining sandy soils known as spodosols or by the local name of *kerangas,* and peaty soils known as *kerapah.* Also in the category of fragile soils fall the peatswamp soils of coastal-deltaic and upland terraces. Schmidt-Lorenz (1986) gives a pantropical synopsis of tropical soils and their nomenclature in the various systems of classification, and Burnham gives a regional review for Southeast Asia in Whitmore 1986. The kerangas and kerapah soils differ in some ecologically important features from the more typical clay-loam soils (ultisols, oxisols or acrisols, and feralsols) found under "climatic climax" mixed dipterocarp forest characteristic of the region (Dames 1956, 1962). Kerangas and kerapah soils, which are generally oligotrophic (nutrient deficient) but diverse in texture and depth (Bruenig 1966, 1974; Bruenig and Schmidt-Lorenz 1985), have the following distinctive features:

1. The surface humus layer is usually fibrous and very acid, with a dense root mat rich in polyphenols (tannins). This surface layer is 10–20 cm thick in kerangas, and may be several meters thick in kerapah.

2. The kerangas horizons in sandy soils vary greatly in depth and are typically composed of uncoated mineral particles with a characteristic pale color, indicating heavy leaching. Below this layer is often a reddish-brown humic or humic-iron pan in which the leached materials have accumulated. The heavy clay soils are dense, bleached to a great depth, and shallow rooted, except for fine roots penetrating along structural clefts.

3. Internal water drainage is unfavorable—either rapid and excessive

(sands) or impeded (bleached clay soils and sands over impervious pan or subsoil). Rapid lateral water flow and extreme fluctuations of soil moisture often occur.

4. These soils often discharge a tea-colored, strongly humic "black-water."

The peatswamp soils of Sarawak are geomorphically diverse (Anderson 1961a, 1964a, 1983). The surface of the peatswamp may be flat or dome-shaped, and the substrate is varied in topography and nature (fig. 3.1). Anderson suspected that permanent waterlogging at the periphery of a swamp and periodic or episodic drought in the center of a swamp were major factors determining local ecological conditions. The distribution of complex organic compounds and concentrations of major and trace minerals may vary among and within peatswamps, and these minerals may also influence the abundance of such economically important timber species as ramin (*Gonystylus bancanus* [Miq.] Kurz) (Anderson 1961a; Bruenig and Sander 1983).

VEGETATION

The kerangas and kerapah communities are floristically heterogeneous (Newbery 1991), in accord with the heterogeneous site conditions (see figs. 3.1, 3.2; Bruenig 1976). Typically, these communities have a lower canopy and are less species rich than mixed dipterocarp forest. The small, leathery leaves in the aerodynamically smooth canopy are morphologically adapted to drought, as in the Amazonian Evergreen Caatinga forest (fig. 3.3; Bruenig 1966, 1970; Bruenig, Alder, and Smith 1979). The peculiar structural, morphological, and biochemical features of the leaves can be interpreted as adaptations to the stresses from droughts, nutrient deficiencies, and possibly organic and inorganic toxins in the soil (fig. 3.4; Bruenig 1966, 1971, 1973b; Proctor et al. 1983).

The peatswamp forest has fewer species and is much more homogeneous than the kerangas and kerapah forests (fig. 3.5). It can be subdivided into six distinct phasic communities (Anderson 1961a, 1963, 1983). The initial phasic community (PC1) of the primary peatswamp forest is the ramin (*Gonystylus bancanus* [Miq.] Kurz) mixed forest, which occupies extensive tracts of newly forming peatswamp at the coastal-deltaic fringe, or narrow belts at the fringes of older peat domes. Ramin forest, structurally the most complex and species-rich phase of the peatswamp forests, covers about 80% of the peatswamp area in Sarawak. The next two phasic communities are dominated by alan (*Shorea albida* Sym.). Ramin forest is gradually pushed out by the vigorous and advancing alan forest (Bruenig and Sander 1983). In the second phase, PC2, the association of *Shorea albida* with *Gonystylus bancanus* and *Stemonurus secundiflorus*,

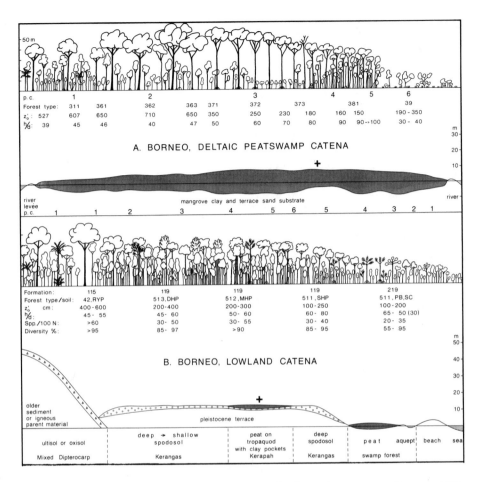

FIGURE 3.1   The sequence of (*A*) phasic communities (PCs) 1 to 6 in peatswamp forests and (*B*) in lowland mixed dipterocarp forests (forest type 42) adjoining the kerangas (513, 512, 511) and kerapah (511 on peat bog) catena on a Pleistocene terrace. The circles indicate predominately sandy soils, the triangles predominantly low-activity clayey loam soils, and the shaded areas peaty soils. *RYP* = red yellow podzolic soil (acrisol/ferrasol/ultisol/oxisol); *HP* = deep, medium deep, and shallow spodosol (humus podzol) on terrace sediment; *PB* = stagnic peat bog. Near the beaches true tropaquod (groundwater podzol) is found. Forest types are defined according to Bruenig 1969, 1976. *Spp./100N* = number of tree species in 100 individuals greater than 1 cm diameter at breast height. The vegetation profiles are at a larger scale than the site-soil profiles. The two crosses on the site transects mark the endpoints of the vegetation profiles.

known as alan forest, usually forms a narrow band at the flank of the peat dome but may cover large tracts in some flat areas with less-developed dome shape. Gaps of 0.1 to 0.3 ha, attributed to lightning and wind, are conspicuous and common in the aerodynamically rough canopy (Bruenig 1973a). The third phasic community (PC3), alan bunga forests (a *Shorea albida* consociation), follows

FIGURE 3.2 Simple kerangas forest on shallow, very poor soils at Samunsam Wildlife Sanctuary, Sarawak. Note the high density of small trees. Photograph supplied by E. F. Bruenig.

next on the flank of the peat dome. This forest has a dense, uniform, tall, and aerodynamically moderately rough-to-smooth canopy. Lightning gaps are smaller than in PC2, but windthrow gaps are larger and more common. The defoliating hairy caterpillar has destroyed much of this forest type since 1947, but the damage apparently stopped in the 1970s (Anderson 1961b, 1964b, Bruenig 1973a).

The next phasic communities, PC4, PC5, and PC6, are forest, woodland, and scrub, respectively. These communities are species-poor, and the plants often have leathery leaves, are low in stature and biomass, and probably grow slowly. PC5 and PC6 are floristically and physiognomically more closely related to kerangas and kerapah than to PC1 (Bruenig 1966). The unfavorable growth conditions at the center of the peat dome, the fragility and low commercial value of the growing stock, and the uniqueness of flora and ecosystem make any commercial use undesirable.

PRODUCTION AND PRODUCTIVITY

The net aboveground primary productivity in kerangas, kerapah, and peatswamp forests is probably considerably lower than in the mixed dipterocarp forest. The annual rates of net primary production (NPP) and of marketable tim-

FIGURE 3.3   Very poor, relatively low kerangas forest with dense, uniform, and smooth canopy, alternating with the large-crowned, aerodynamically rough and diverse mixed dipterocarp forest on the Usun Apau plateau in the interior of Sarawak. Photograph supplied by E. F. Bruenig.

ber are little known but may be deduced from the height of the canopy. NPP/ha in closed pristine forest during the building phase has been estimated to range from 5 tons dry matter (see fig. 3.1, right) to 40 t d.m. (see fig. 3.1, left) (Bruenig 1966, 1974).

In kerangas, higher-growing stock densities and tall canopies are associated with higher rates of growth of biomass and commercial timber in forests with the commercially important timber trees bindang (*Agathis* spp.), alan, and ramin. In the peatswamp and kerapah forests, this relation between timber stocks and growth rates does not exist, even when commercial timbers

FIGURE 3.4   Roots above the ground surface in kerangas forest grow into decaying standing and fallen deadwood, into hollow trees, and on the bark surface of standing live trees, probably to scavenge water and nutrients. Such extensive rooting may indicate an extreme reaction to nutrient and water deficiency. In this photograph roots of rhu (*Casuarina nobilis* Whitm., seen on the left) grow upward and downward on the bark of a neighboring bintangor (*Calophyllum* sp.) in Sabal Forest Reserve, Sarawak. Photograph supplied by E. F. Bruenig.

are present. As a result of this low productivity, at least 10% of the peatswamp forests and more than 50% of the kerangas and kerapah forests are unsuitable for production forestry. When trees are commercially harvested from these forests, the subsequent growth rates and regeneration of trees are often disappointingly low.

DEAD BIOMASS: CRUCIAL FOR SUSTAINABILITY

The only effective adsorptive complexes and sources for nutrients in most kerangas and kerapah forests and in all peatswamp forests, except for atmospheric inputs, are the decaying standing trees, coarse and fine litter, and organic matter in the soil. In pristine tall kerangas in Research Plot (RP) 146 in Sabal Forest Reserve on medium deep sandy podzol, the mean volume and biomass of aboveground standing dead trees greater than 10 cm diameter in 1989–90 were 9.4 m^3 and 7.5 t ha^{-1} d.m., respectively. Fallen trees and branches amounted to 96.5 m^3 and 63 t respectively. The total deadwood biomass was 70.5 t ha^{-1}, corresponding to 15.8% of the 447 t ha^{-1} of aboveground living biomass (Droste 1992). This value is in close accord with the data from Amazonian Caatinga at San Carlos de Rio Negro (Alder et al. 1979; Bruenig, Alder, and Smith 1979).

The biomass of dead and dying trees, rotting wood and hollow trunks, fine and coarse litter, and soil organic matter in kerangas has several

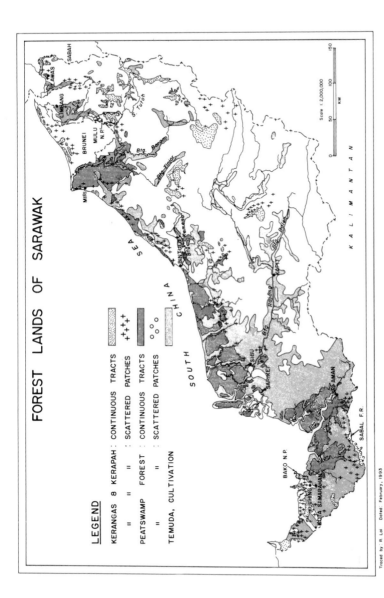

FIGURE 3.5 Distribution of kerangas, kerapah, peatswamp, and cultivated lands in Sarawak. Kerangas occurs in large continuous areas on terrace landscapes, hard siliceous sandstones, and some ultrabasic materi- als, as well as scattered on perched terrace remnants, ridge tops, and dip slopes on sandstone-derived soils. Map supplied by the Sarawak Forest Department.

ecological functions: it provides a direct and indirect source of nutrients for the living trees, supplies food for ground and soil organisms (animals, plants, microbes), improves rootability by adding material that stabilizes the soil, and provides an adsorption complex for water and mineral nutrients. Exposure and disturbance of the soil by excessively heavy and careless logging in kerangas, kerapah, and peatswamp increase soil temperature and biological activity, consequently accelerating the decomposition (respiration) of the litter and soil organic matter. Slash-and-burn conversion of kerangas to tree plantations immediately destroys most of the deadwood component and surface humus. This destruction of the organic matter, combined with full exposure of the soil to rain, sunlight, and wind, accelerates the process of soil degradation.

PRINCIPLES FOR HARVESTING AND SUSTAINABLE MANAGEMENT

Kerapah forests and the peatswamp forest of PC4, PC5, and PC6 are economically so poor (in terms of the number and size of trees) and ecologically so fragile that they are unsuitable for sustainable production forestry. The same applies to kerangas forests on all types of shallow gley-type clays and shallow sandy podzol soils, and on slopes greater than 25%. Alan peatswamp forest of PC3 has been harvested by clear-felling since the 1950s, but natural regeneration has failed. The persistent failure of *Shorea albida* to regenerate in areas damaged by insect attack (for details see Anderson 1961b) and in the commercial clear-cut areas is symptomatic of a fragile regeneration regime that is still a complete mystery to foresters and ecologists. Clear-felling in the present manner obviously removes an important resource and damages the ecosystem without a promise of restoration. Planting, though possible, is risky and expensive (Lai 1978). A shelterwood, shelterwood-strip, or shelterwood-patch system would seem to be an economically viable option, but it has never been tried.

The alan batu and ramin phasic communities are theoretically easy to manage by a selection-silvicultural system with long felling cycles and a high diameter limit. The natural phasic dynamics of PC2 favor alan, and ramin thus cannot establish itself or be introduced because of toxins in the alan peat (Bruenig and Sander 1983). In PC1 the natural regeneration practice for the major timber species is well established (Chai 1986; Lee 1979, 1991). The major management problems are the natural ecosystem dynamics, which drive out ramin, and the unreliable and often poor response of intermediate ramin trees and seedlings to canopy opening (fig. 3.6). More comprehensive, system-analytical growth studies are urgently needed (Chai 1991). In the meantime, mounting pressures for premature reentry of logged areas and reduced diameter

FIGURE 3.6 Selective logging has been practiced successfully for many decades in peatswamp forest. Because heavy machinery cannot be used on soft peatswamp soils, logging damage is minimal. However, post-felling regeneration of certain important timber species, such as ramin, is often poor. Photograph by R. Primack.

limits for harvesting must be resisted to prevent lasting damage to this fragile ecosystem.

Kerangas forests on the more favorable soils (see fig. 3.1, left) can be commercially used and sustainably managed under a regeneration-oriented selection-silviculture management system. Biodiversity and canopy preservation are essential for sustainability. The key to sustainable logging is low-impact logging, which incorporates mandatory tree marking, directional felling, optimizing of layout and construction of extraction lines, low intensity of cutting, and compliance with the ITTO Guidelines for the Sustainable Management of Natural Tropical Forest (ITTO 1990, 1991, 1992a, 1992b; Tropical Forest Initiative 1993). In addition to bindang (*Agathis borneensis* Warb.), sempilor (*Dacrydium* spp.), and a variety of high-quality timbers from dipterocarp species mainly of the genera *Shorea, Dipterocarpus, Vatica,* and *Anisoptera,* a wide range of specialty timbers can be produced. Other economically promising options in forest management include reestablishment of the almost extinct species gaharu (*Aetoxylon sympetalum* [Steen. ex Domke] Airy Shaw), valued for incense production, and planting of fine-grade rattan species after logging. Extraction of nontimber resources in peatswamp and kerangas forests by local people is now greatly reduced because of exhaustion of resources, substitution by other materials, and collapse of the markets.

CONCLUSIONS

Until well into the 1970s, the catchwords of tropical forestry were natural forest uniformization and fast-growing plantations. Finally, however, people are recognizing the risks these operations pose to long-term sustainability and are also appreciating the interrelated ecological, economic, and social values of biological diversity. The keystone role of the upper canopy, formed by the structural species *sensu* (Solbrig 1991a, 1991b), for biodiversity and sustainability is now better understood. Harvesting and management in natural and planted tropical forest on all sites, but especially on high-risk fragile sites, must maintain high levels of biodiversity within and between stands to secure sustainability (ITTO 1990, 1991, 1992a, 1992b). The implications for rainforest management on fragile soils are twofold. First, the natural site-specific species richness should be maintained, and augmented if possible, to ensure adequate ecological and economic viability. This excludes uniformization, clear-cutting of the stand, and conversion to plantations, and calls instead for natural forest management. Restoration of deforested sites on fragile soils should be left to natural succession, which may be enhanced by planting after the soil has been sufficiently improved. Second, practical, site-specific criteria and indicators are required to assess and continuously monitor biodiversity and canopy condition for management purposes (Bruenig and Schneider 1992).

The first condition requires low-impact logging, optimally planned and established access systems, and patchy opening of the canopy strictly within the limits of natural gap formation to which the ecosystem is adapted. Although the basic sustainability principles that apply in these forests are the same as in mixed dipterocarp forests (Appanah et al. 1990; Appanah and Weinland 1991), additional restrictions are imposed in kerangas forest because of the higher risk of fire and soil degradation. Total protection is the only acceptable option in the poorer types of peatswamp (PCs 4, 5, and 6) and in the more extreme kerangas and kerapah forests, and very cautious low-impact naturalistic silvicultural management is the only sustainable option for the more favorable sites.

The second condition requires the development of local, site-specific standards based on ITTO criteria. Effective application of these criteria must include the use of remote sensing to monitor the condition of the main canopy and the quality and extent of the access and extraction system. Continuous ground control and selected spot-checks should occur at points identified as critical (Bruenig and Schneider 1992). In this way, the canopy, soil, and adjacent water bodies, the most important elements of sustainable forest management on fragile soils, can be maintained in a condition compatible with the principles of sustainability.

ACKNOWLEDGMENTS

The authors thank the Sarawak Forest Department, which has supported scientific and experimental work and the development of management systems for the past forty years. Additional support for scientific research has been given by the German Research Foundation (DFG) and the Volkswagen Foundation, Germany. The enumeration and soils survey of RP 146 have been a cooperative project with the Forest Research Branch of the Sarawak Forest Department. Julia Philip, of the Forest Department, processed the manuscript.

REFERENCES

Alder, D., E. F. Bruenig, J. Heuveldop, and J. Smith. 1979. Struktur and Funktionen in Regenwald des Internationalen Amazon-Oekosystem-Projektes. *Amazoniana* 6: 423–444.

Anderson, J. A. R. 1961a. The ecology and forest types of the peatswamp forests of Sarawak and Brunei in relation to their silviculture. Ph.D. diss., University of Edinburgh, Edinburgh.

———. 1961b. The destruction of *Shorea albida* forest by an identified insect. *Empire For. Rev.* 40, 103: 19–28.

———. 1963. The flora of the peatswamp forest of Sarawak and Brunei, including a catalogue of all recorded species. *Gard. Bull. Singapore* 20: 131–228.

———. 1964a. The structure and development of the peatswamps of Sarawak and Brunei. *J. Trop. Geogr.* 18: 7–16.

———. 1983. The tropical peatswamps of western Malesia. In *Mires: Swamp, Bog, Fen and Moor, Part B: Ecosystems of the World 4B.* A. J. P. Gore, editor. Elsevier, Amsterdam.

Anderson, J. A. R., and J. Muller. 1975. Palynological study of a holocene peat and a miocene coal deposit from northwest Borneo. *Rev. Palaeobot. Palynol.* 19: 291–351.

Appanah, S., and G. Weinland. 1991. Will the management systems for hill Dipterocarp forests stand up? *J. Trop. For. Sci.* 3, 2: 140–158.

Appanah, S., G. Weinland, H. Bossel, and H. Krieger. 1990. Are tropical rainforests non-renewable? An enquiry through modelling. *J. Trop. For. Sci.* 2, 4: 331–348.

Bruenig, E. F. 1966. Der Heidewald von Sarawak und Brunei eine Studie seiner Vegetation und Okologie. Habilitation Thesis, University of Hamburg, Hamburg Publ. Mitt. BFH No. 68, 431.

———. 1969. Forest classification in Sarawak. *Malay. For.* 32, 2: 143–179.

———. 1970. Stand structure, physiognomy and environmental factors in some lowland forests in Sarawak. *J. Trop. Ecol.* 11, 1: 26–43.

———. 1971. On the ecological significance of drought in the equatorial wet evergreen (rain) forest of Sarawak (Borneo). In *Transactions of the First Aberdeen-Hull Symposium on Malesian Ecology.* Univ. Hull, Dept. Geogr., Misc. Ser. 11.

———. 1973a. Some further evidence on the amount of damage attributed to lightning and windthrow in *Shorea albida* forest in Sarawak. *Commonw. For. Rev.* 52, 153: 260–265.

———. 1973b. Species richness and stand diversity in relation to site and succession in forests in Sarawak and Brunei. *Amazoniana* 4: 293–320.

———. 1974. *Ecological Studies in Kerangas Forests of Sarawak and Brunei.* Borneo Literature Bureau for Sarawak Forest Department, Kuching.

————. 1976. Classifying for mapping of kerangas and peatswamp forest examples of primary forest types in Sarawak/Borneo. In *The Classification and Mapping of South-East Asian Ecosystems*. P. S. Ashton, editor. Univ. Hull, Dept. Geogr., Misc. Ser. 17.

Bruenig, E. F., D. Alder, and J. P. Smith. 1979. The international MAB Amazon rainforest ecosystem pilot project at San Carlos de Rio Negro: Vegetation classification and structure. In *Transactions of the Second International MAB-IUFRO Workshop on Tropical Rainforest Ecosystems Research*. S. Adisoemarto and E. F. Bruenig, editors. Chair of World Forestry, Hamburg-Reinbeck, Spec. Rep. No. 2.

Bruenig, E. F., and H. Klinge. 1969. Forestry on tropical podzols and related soils. *J. Trop. Ecol.* 10, 1: 45–58.

Bruenig, E. F., and N. Sander. 1983. Ecosystem structure and functioning: Some interactions of relevance to agroforestry. In *Proceedings of a Consultative Meeting, Nairobi, Kenya*. P. A. Huxley, editor. ICRAF, Nairobi.

Bruenig, E. F., and R. Schmidt-Lorenz. 1985. Some observations on the humic matter in Kerangas and Caatinga soils with respect to their role as sink and source of carbon in the face of sporadic episodic events. *Mitt. Geol.-Palaeontol. Inst. Univ. Hamburg, SCOPE/UNEP Sonderband* 58: 107–122.

Bruenig, E. F., and T. W. Schneider. 1992. Assessing and monitoring biological diversity in sustainable management of tropical rainforest. Bangkok, Princes Congress II, Chulabhorn Res. Inst., Mahidol University.

Chai, F. Y. C. 1986. Silvicultural treatment in Mixed Swamp forest. Paper presented at the Ninth Malaysian Forestry Conference, October. Sarawak Forest Department, Kuching.

————. 1991. Diameter increment models for the Mixed Swamp forests of Sarawak. M.Sc. thesis, Department of Forestry, University of British Columbia, Vancouver, British Columbia, pp. xii, 117.

Dames, T. W. G. 1956. Preliminary classification of some of the soils of Sarawak and Brunei. Misc. Rep. For. Dep., Kuching. Mimeograph, 9.

————. 1962. *Soil Research in the Economic Development of Sarawak*. FAO/EPTA Rep. No. 1512, Rome.

Droste, H. J. 1992. Erste Ergebnisse einer Totholzanalyse in Sabal Forest Reserve (First results of a dead wood analysis in RP 146, Sabal Forest Reserve). Hamburg, report, 19.

ITTO (International Tropical Timber Organization). 1990. ITTO guidelines for the sustainable management of natural tropical forests. Yokohama, International Tropical Timber Organization, ITTO Policy Devel. Ser. No. 1.

————. 1991. Beyond the guidelines: An action program for sustainable management of tropical forests. Yokohama, International Tropical Timber Organization, *ITTO Techn. Ser.* No. 7, pp. vi, 187, app. p. 6.

————. 1992a. Criteria for the measurement of sustainable tropical forest management. Yokohama, International Tropical Timber Organization, *ITTO Policy Dev. Ser.* No 3, pp. i, 5.

————. 1992b. ITTO guidelines for the establishment and sustainable management of planted tropical forests. Yokohama International Tropical Timber Organization. *ITTO Policy Dev. Ser.* No. 4, pp. iv, 38.

Klinge, H. 1968. Tropical podzols. FAO Report MR/69832, FAO/UN, Rome.

Lai, K. K. 1978. Line planting trials of *Shorea albida*. Sarawak Forest Department, For. Res. Rep. SR 21.

Lee, H. S. 1979. Natural regeneration and reforestation in the peatswamp forests. Paper presented at the Symposium on Silvicultural Technologies, Tokyo, November 1978. Trop. Agric. Ser. No. 12.

————. 1991. Utilization or conservation of peatswamp forests in Sarawak. Paper presented at the International Symposium on Tropical Peatland, Kuching. Newbery, D. McC. 1991. Floristic variation within Kerangas (heath) forest: Re-evaluation of data from Sarawak and Brunei. *Vegetatio* 96: 43–86.

Proctor, J., J. M. Anderson, P. Chai, and H. W. Vallack. 1983. Ecological studies in four contrasting lowland rain forest in Gunung Mulu National Park, Sarawak, I: Forest environment, structure and floristics. *J. Ecol.* 71: 237–260.

Richards, P. W. 1952, 1989. *The Tropical Rain Forest.* Cambridge University Press, Cambridge.

Schmidt-Lorenz, R. 1986. Die Boeden der Tropen und Subtropen. In *Handbuch der Landwirtschaft und Ernaehrung in den Entwicklungslaendern.* P. V. Blanckenburg and H. D. Cremer, editors. Ulmer Verlag, Stuttgart.

Solbrig, O. 1991a. Biodiversity-scientific issues and collaborative research proposals. Paris, Unesco, MAB *Digest* No. 9, p. 77.

Solbrig, O., editor. 1991b. *From Genes to Ecosystems: A Research Agenda for Biodiversity.* International Union of Biological Sciences, Cambridge, Mass., and Paris.

Tropical Forest Initiative. 1993. Kriterien zur Beurteilung einer nachhaltigen Bewirtschaftung tropischer Wälder (Criteria for the assessment and valuation of concession enterprises in the tropical humid forest zone). Initiative Tropenwald, Berlin. Cyclostyled draft (fax 30–27822725).

Whitmore, T. C. 1986. *Tropical Rainforests of the Far East.* Oxford University Press, Oxford.

4

LINKS BETWEEN VERTEBRATES AND THE
CONSERVATION OF SOUTHEAST ASIAN RAINFORESTS
Junaidi Payne

Vertebrates—mammals, birds, reptiles, amphibians, and fish—are a significant
component of rainforest ecosystems. Most of the native vertebrate species of
tropical rainforests have evolved with or in forest habitats, so their survival is
closely tied to that of the forests. Borneo, for example, has more than 200 mammal species, including bats, which are believed to require forest or tree cover for
their long-term survival as breeding populations. The island also has about 350
species of birds, more than 200 species of reptiles, and more than 80 species of
amphibians, all of which are forest dependent. The number of species represented as breeding populations within a particular forest area, however, varies
considerably from place to place.

Not only are many vertebrate species dependent on rainforests for
their survival, but all participate in the cycling of nutrients and contribute in
various, often subtle, ways to the maintenance of forest integrity. Some contribute directly to the survival and evolution of ecosystems by pollinating flowers and dispersing seeds. This chapter outlines the roles of vertebrates in the
rainforest ecosystem, concentrating on examples from Malaysia and Borneo,
and then addresses some wildlife conservation issues for which the support of
forest managers and policymakers is now critical.

POLLINATION AND SEED DISPERSAL BY VERTEBRATES
IN SOUTHEAST ASIAN RAINFORESTS

Insects are the major pollinators of Southeast Asian rainforest plants (Appanah
1985), but nectar-feeding birds, notably flowerpeckers (Dicaeidae), sunbirds,
and spiderhunters (Nectariniidae), are also significant pollinators. Among
mammals, the most important pollinators, in terms of number of animals involved and range of plants pollinated, are the fruit bats (Pteropodidae). At least
306 plant species have well-documented relations with these bats, which act as
pollinators, seed dispersers, or both (Fujita 1988). At least 450 economic prod-

ucts, including timber, fruits, fuelwood, fiber, medicines, tannins, and dyes, are derived from plants for which fruit bats are pollinators or dispersal agents. Fruit bats of the genera *Eonycteris* and *Macroglossus* pollinate many economically important Southeast Asian trees, including durian (*Durio zibethinus*), petai (*Parkia speciosa*, also *P. javanica*), jambu (*Eugenia malaccensis*), mango (*Mangifera* spp.), banana (*Musa* spp.), and bakau (*Rhizophora* spp.) (Gould 1978; Start and Marshall 1976).

Seed Dispersal

Numerous examples of dispersal of rainforest plant seeds by vertebrates have been reported, but most are documented only as brief notes within studies of other aspects of plant or animal biology. Ridley's reviews (1894, 1930) have not been superseded for Southeast Asian forests. Despite the enormous increase of ecological studies in tropical rainforests, the main dispersing agents of most plants in Southeast Asian rainforests remain unknown. The observations that do exist are largely qualitative or anecdotal, lacking documentation of the number of seeds involved and their ability to germinate. Seeds of relatively few species, mainly tall trees and climbers (such as Dipterocarpaceae, *Koompassia* spp., and *Alstonia* spp.), are dispersed without the aid of animals, mainly through gravity and wind. A study in Ghana (Lieberman and Lieberman 1986) showed that, for most woody rainforest plants, successful germination is not completely dependent on dispersal of seeds by fruit-eating animals. Nevertheless, the dispersion of trees through tropical rainforests indicates that a proportion of the trees must have been derived from seeds that were actively transported away from the parent by fruit eaters.

Treeshrews (Tupaiidae), once regarded as insectivores, now are known to feed on a variety of fruits. In Kinabalu Park, Sabah, the large treeshrew (*Tupaia tana*) and the plantain squirrel (*Callosciurus notatus*) have been observed feeding on the fruit of *Rafflesia keithii*, a member of the genus of plants known for enormous flowers (Emmons, Jamili, and Alim 1991). It seems likely that these and other small mammals disperse the seeds of this unusual plant, which is parasitic on vines of the genus *Tetrastigma*. Treeshrews at Danum Valley and Poring in Sabah were observed feeding on fruits and spitting out the seeds of wild fig (*Ficus*) species and of Sapindaceae and Myrsinaceae species (Emmons, personal communication, August 1989).

Throughout Southeast Asia, large fruit bats of the genus *Pteropus* (fig. 4.1), known as flying foxes, are often perceived as destroyers of orchard fruits. Both these and smaller fruit bats, however, disperse seeds of wild and planted trees by carrying fruits away from the parent tree to their roosting sites,

FIGURE 4.1   Large, wide-ranging fruit bats, also known as flying foxes, feed on fruits in orchards, causing considerable damage, but they are important pollinators and seed dispersers for wild trees in a variety of forest habitats. Photograph by the Sarawak Forest Department.

where they consume the flesh and drop the seeds (Cox et al. 1991). Many people regard some primates as seed dispersers and others as seed predators, but in fact all primates and most other vertebrates that feed on fruits probably disperse some seeds and destroy others. Galdikas (1982), for example, noted that in feces of wild orangutans (*Pongo pygmaeus*, fig. 4.2) some seeds are intact while other seeds of the same species are broken or destroyed. During a six-month period in Tanjung Puting Reserve in southern Borneo, 94% of orangutan feces were found to contain seeds representing at least 23 woody plant species. Many plant species are dispersed by both fruit bats and primates, including tree species that are commercially valuable (such as some *Diospyros* species, members of the ebony family) or may become valuable (such as *Neolamarckia cadamba*, which regenerates well in heavily disturbed lowland forests).

Rodents are typically seed predators, yet they may disperse viable seeds in some circumstances (when transporting and burying seeds for future consumption, for example, or when carrying fruits to young animals in the nest). Malaysian tree squirrels (Sciuridae) have been observed transporting fruits of various forms and sizes, including *Alphonsea elliptica* and *Aglaia* species, through the forest canopy (Becker, Leighton, and Payne 1985; Payne 1979). Some of these fruits may be dropped away from the parent tree. Porcupines

FIGURE 4.2 Orangutan populations have declined dramatically, reducing the animals' important role as seed dispersers. Photograph by the Sarawak Forest Department.

(Hystricidae) appear to move seeds of the Borneo ironwood tree (*Eusideroxylon zwageri*) from under the parent tree and at least partially eat them elsewhere (Payne, Francis, and Phillipps 1985).

Some civet species (Viverridae) have a mixed diet that includes animal and plant materials. Civet feces containing seeds of trees and woody climbers (such as *Gnetum* spp. and Annonaceae) are commonly found on exposed surfaces in Southeast Asian rainforests. Two Southeast Asian civet species in particular, the common palm civet (*Paradoxurus hermaphroditus*) and the masked palm civet (*Paguma larvata*), are reported to disperse the seeds of a variety of plant species (Medway 1983). Rabinowitz and Walker (1991) found that 12.4% of food items in a collection of civet feces in dry tropical forest in Thailand were seeds, representing at least eighteen plant species.

The role of large terrestrial mammals in dispersing rainforest seeds has not been fully investigated in any Southeast Asian region. Available observations suggest that wild pigs (Suidae) and deer (Cervidae) either grind and digest seeds or feed on the fleshy parts of fruits in situ, leaving the seeds under or near the parent tree. For many plant species, the appearance and taste of the fruit (commonly hard, green, odorless, and bitter or astringent) do not indicate the usual mode of dispersal, but large terrestrial mammals would seem to be among the candidates. Asian rhinoceroses feed mainly on leaves but include fallen fruits in their diet. For example, the Javan rhinoceros (*Rhinoceros sondaicus*) is reported to disperse seeds of *Pandanus* species and *Dillenia aurea* (Ammann 1985). Two seeds recovered from the stomach of a wild-caught Asian two-horned rhinoceros (*Dicerorhinus sumatrensis*), which died in the London Zoo about a century ago, were identified as those of *Mezzettia leptopoda* (J. A. R. Anderson, personal communication, May 1984). The fruit of this Southeast Asian evergreen rainforest tree is about 10 cm wide and very hard, and has a resinous odor. In Ghana, traces of fruits were found in 93% of dung piles of forest-dwelling elephants (Short 1981). The rarity of the largest plant-eating mammals (tapirs, rhinos, wild cattle, and elephants) in most accessible areas makes research difficult, but observations of hunter-naturalists from times when these animals were common provide valuable insights. Hubback (1939) indicates that the Asian two-horned rhinoceros favors the fallen fruits of wild mangoes (*Mangifera* species). The diet of wild cattle (Bovidae) consists mainly of leafy material but evidently includes fallen fruits; the cattle disperse viable seeds. For example, Foenander (1952) indicated that the now-endangered seladang (*Bos gaurus*) disperses seeds of legume trees.

Many Southeast Asian birds are seed dispersers. In contrast to American tropical forests, however, where specialist fruit eaters form a large component of the bird community, in Southeast Asia most fruit-eating birds also feed on other items, such as insects and nectar. One of the few fruit-eating specialists of Malaysia is the green broadbill (*Calyptomena viridis*), which disperses seeds of understory trees and palms with large seeds. These birds decline in number after logging of dipterocarp forests (Lambert 1990). All 8 hornbill species (Bucerotidae) in Borneo, some of which may travel great distances, include fruits in their diet, which features plants of the families Moraceae, Lauraceae, and Meliaceae (Becker and Wong 1985; Kemp and Kemp 1974; Leighton and Leighton 1983). At least 15 species of pigeons and doves (Columbidae) of Borneo and offshore islands feed on fruits, and most or all act as seed dispersers. Of the 24 species of bulbuls (Pycnonotidae) found in Borneo, at least 14 feed on fruits (Smythies 1981). Observations by field biologists suggest that bulbuls play a role in introducing the seeds of some pioneer trees into deforested areas. Some of the small nectar- and insect-eating birds also disperse small seeds. For example, the purple-naped sunbird (*Hypogramma hypogrammicum*) disperses seeds of *Poikilospermum suaviolens*, a climbing plant often present in secondary growth and forest edges (Lambert 1991).

Many vertebrates of the Southeast Asian rainforests do not have highly specialized dietary requirements. Few are confined to specific forest types; however, close coevolutionary relations do exist between some plants and animals. For example, fig plants (*Ficus* spp.; Moraceae), especially large strangling figs, play a disproportionately large role in sustaining the vertebrate community, considering their abundance in the rainforest ecosystem. The fruits are eaten by almost all species of mammals and birds that include rainforest fruits in their diet. At one lowland dipterocarp forest study site in Peninsular Malaysia, at least 60 of more than 200 resident bird species have been observed to feed on fig fruits (Lambert 1989). The seeds of strangling figs must be dispersed by animals in order to germinate and grow on the host tree. Large families of woody plants that exhibit many close, mutually beneficial relations with vertebrates (the plants provide food for animals, and pollination or seed dispersal is carried out by those animals) include Annonaceae, Bombacaceae, Burseraceae, Euphorbiaceae, Lauraceae, Meliaceae, Moraceae, and Sapindaceae.

In the seasonally flooded forests of the Amazon basin, seeds of at least 26 tree species are believed to be dispersed partly or mainly by fish (Goulding 1980), but whether any Southeast Asian fish species disperse seeds is unknown. In Sabah, northern Borneo, at least 13 species of freshwater fishes feed on fruits that fall from overhanging trees into watercourses or onto flooded land (Inger and Chin 1962). The massive reductions in Peninsular Malaysian fish population densities in most river systems and complete loss of some species (Mohsin and Ambak 1983) may have affected dispersal of seeds, to an unknown extent.

Other Relations

Trees and climbing plants of the legume family (Fabaceae) are particularly important in Malaysian forests in supplying food—including fruit, leaves, shoots, flowers, and bark—for primates, squirrels, and possibly other mammals, especially when fruits are scarce (Chivers 1980; Marsh and Wilson 1981). The term *keystone* has been used to refer to plant resources, such as figs and legumes, that play this type of prominent role in sustaining plant-eating animals through periods of food scarcity (Terborgh 1986).

Vertebrates may play various indirect roles in maintaining the integrity and stability of tropical rainforests. Primates and ungulates, for example, feed on leaves and leaf shoots, selecting certain plant species and presumably reducing the growth potential and competitive advantage of those species. Some vertebrates open up thick vegetation, which suppresses regeneration of trees, by trampling or by feeding on that vegetation. Elephants (*Elephas maximus*) and orangutans in Sabah, for example, feed on the shoots and soft stems of climbing

bamboo (*Dinochloa* spp.), a plant that invades forest openings and hinders regeneration of timber trees in some areas. Vertebrates that turn over leaf litter
and topsoil in search of food (notably wild pigs) or that burrow (such as terrestrial rodents) may damage existing vegetation but enhance prospects for successful germination of seeds.

Trees with holes or rotten, hollow trunks are common in most rainforests. The holes or spaces are often inhabited by bats, flying squirrels, porcupines, or rats, and birds use some as nesting sites. These circumstances may be
advantageous not only to the animals but also to the trees, which receive nutrients from the animals' waste products (Whitten et al. 1984). The relations
among the entire array of plants and vertebrates in Southeast Asian rainforests
are highly complex, with one animal-plant relation providing the conditions
needed to support other relations. Few specific plant-animal interactions, however, have been investigated in detail.

FACTORS DETERMINING VERTEBRATE SPECIES DISTRIBUTION AND ABUNDANCE

The factors that determine the distribution and abundance of wild vertebrate
populations in tropical rainforests have not been critically examined. Rare
events such as volcanic eruptions, floods, storms, and changes in sea level are
likely to have profound, long-lasting effects on the distribution of vertebrates.
Human activities, notably hunting, traditional farming, and logging, are commonly blamed for the decline of large animals. These activities do influence vertebrate populations (Bennett and Zainuddin, this volume), but they alone
cannot account for the distribution and abundance of large mammals on a regional scale (Payne 1992). Availability of food, which undoubtedly influences
vertebrate populations locally, does not explain the documented distribution
patterns of large animals in Southeast Asia, even before the clearing of extensive
tracts of lowland forests for permanent agriculture and human settlement.
Large mammals were formerly more abundant in these forests, so it is difficult
to determine the natural factors that influenced population density and ranging
of these animals. It is possible only to draw attention to certain correlations.

In general, large mammals in Malaysia are most abundant on the
most mineral-rich soils. An extreme example is the distribution of the proboscis
monkey (*Nasalis larvatus*), a Bornean endemic, which is generally associated
with alluvial soils, possibly because only these soils provide adequate minerals or
adequate amounts of food to sustain breeding populations (Bennett and
Sebastian 1988). Payne (1988) presented evidence that population densities of
wild breeding orangutans in the lowlands of eastern Sabah were relatively con-

stant for about 20 years at locations that remained under forest cover, irrespective of intensity of timber extraction. High population densities of this species in Sabah in fact occur only in freshwater swamp and low-altitude dipterocarp forests. It is not clear, however, which factors, perhaps associated with particular soils, limit the distribution and abundance of wild animal populations.

Availability of fresh water may influence distribution and abundance of animals, especially in seasonal and "dry" tropical forests. In dry tropical Asian forests, for example, tigers are closely associated with rivers (Rabinowitz 1989). Low concentrations or nonavailability of elements needed by vertebrates also may limit the distribution of some species. The occurrence of large multispecies herds of herbivorous mammals in East Africa has been correlated with high concentrations of sodium, magnesium, and phosphorus in food items (McNaughton 1988). Malaysian rainforest leaves often contain concentrations of sodium and phosphorus below those believed to be required by mammals. Leaf eaters such as rhinos thus may need to obtain these minerals from other sources, such as fruits and natural concentrations of mineral-rich soil or water (often known as "salt licks"). Hunters throughout Southeast Asia in the early decades of the twentieth century often referred to links between salt licks and large mammals. The distribution of elephants and rhinos in Borneo may be limited by such salt licks (Davies and Payne 1982; Payne 1992); however, concentrated sources of salt may not necessarily be manifested as salt licks. For example, Mykura (1989) demonstrated that the concentration of sodium in water varies by at least tenfold among streams of similar size and flow conditions within the Maliau Basin in Sabah. The lack of trace minerals (rare elements such as iodine, molybdenum, and selenium, which vertebrates require in very small quantities) may limit the distribution of some species, especially plant eaters (Payne 1988); however, a preliminary study of trace minerals in soil and plants in Sabah (Mokhtar and Payne, unpublished data) suggests that this is unlikely.

Like plants, most vertebrate species occur within a particular altitudinal range. The factors determining altitudinal preferences of animals are not clear, as those birds and mammals confined to mountain ranges are not clearly limited to montane forests. The distribution of many mountain-dwelling animals encompasses the upper range of hill dipterocarp forests and the lower range of nondipterocarp montane forests. These species may depend on resources provided by both forest types. In contrast, individuals of some lowland species are often found at high altitudes. In these cases, the evidence suggests that lowland forest is essential for breeding populations, and nonbreeding individuals (often immature individuals and mature males) spill over into higher-altitude forests. The implications for conservation are important: the presence of a particular animal species in a reserve or park in a hill range does not necessarily indicate that it can survive in that area as a breeding population. For example, proboscis monkeys and orangutans have been reported in the Maliau Basin Conservation Area in Sabah, but this area does not support breeding populations of either species.

CONSERVATION ISSUES

The identification and establishment of a permanent forest estate (PFE: the forestland allocated permanently for forestry purposes, including protection, within a country or state) are a foundation for timber production, environmental protection, and wildlife conservation. Typically, PFEs are dominated by hill ranges and mountains, with some lowland areas unsuitable for agriculture. These PFEs are inadequate to conserve species that require lowland forests on the more fertile soils—forestland must be reserved where rare species are naturally abundant. Translocation of vertebrates from lowland forests being cleared for agriculture into reserved hill forests is, in most cases, an expensive waste of time and effort (Caldecott and Kavanagh 1983). Populations of mammals and birds are able to survive in lowland forests under current regimes of management for timber production as long as the level of hunting does not increase (Davies and Payne 1982; Johns 1989; Lambert 1990; Payne 1988). Thus, timber forests in the lowlands can contribute significantly to vertebrate conservation. Both the production and protection components of PFEs can be managed effectively, however, only with adequate funds, human resources, and technical expertise.

A PFE will not be adequate for conservation of wild vertebrates, or biological diversity in general, if the traditional foundation of Southeast Asian evergreen rainforest management, involving removal of large trees and forest regrowth through natural regeneration, is replaced by new policies designed to produce high annual levels of timber. Increased intensity of wood extraction from natural forests would profoundly alter the rainforest ecosystem. Areas of concern for vertebrates include removal of trees with holes (nesting sites for hornbills and other wildlife), harvesting of nontraditional tree species (vertebrate food sources), removal of potential keystone plants (figs and legumes, for example), and short cycles of timber extraction from logged forests (leading to loss of microhabitats for sensitive species).

Plantations of trees for wood production are not conservationally equivalent to natural forests. Studies of mammals and birds in tree plantations in Sabah have determined that breeding populations of most species in nearby natural forests are absent (Duff, Hall, and Marsh 1984; Sheldon 1986; Stuebing and Gasis 1989). Vertebrate conservation strategies thus may not be able to accommodate higher-intensity harvesting or extensive tree plantations.

Until 1983, many people believed that rainforests could not be significantly affected by fire. A long dry spell in 1982–83, coupled with unprecedented road access, a spreading human population, and massive amounts of deadwood in logged forests, led to widespread forest fires in northern and eastern Borneo (Beaman et al. 1985; Mackie 1984). Repeated fires in Bornean evergreen forests changed the character of those forests markedly. Vertebrate carnivores in dry dipterocarp forests of Thailand, a habitat maintained by repeated forest fires, can survive only temporarily and only where evergreen forest is present (Rabinowitz 1990). Because of the threat fire poses to critical habitats, forest departments and logging companies need to implement practices that minimize the potential for postlogging forest fires.

Captive breeding and animal translocation programs may be using large proportions of resources made available for wildlife conservation. In some areas, laws concerning hunting and keeping wild animals are strictly enforced, irrespective of whether these practices have any bearing on the survival of wild populations. Both government and nongovernment agencies need to devote greater efforts to a few priorities relating to in situ wild animal populations, such as identification of truly threatened species, investigation of the natural resources that limit their wild populations, and formation of appropriate extensions of PFEs to incorporate additional lowland habitats.

Research to address these concerns needs to cover several areas. First, the remaining forestland supporting permanent breeding populations of rare and endangered vertebrate species at high densities needs to be specifically identified, along with the critical factors that limit wild breeding populations of these species. Second, the roles of vertebrates in pollination and seed dispersal should be investigated in detail. Taxa likely to be of particular interest include fruit bats and small fruit-eating birds such as bulbuls, which may be important in regeneration of logged, burned, and degraded forests and deforested land, as well as rodents and large terrestrial herbivores. Third, the effects of fire on forest regeneration and wild animal populations in the evergreen forest regions merit investigation.

As for policy, all countries and states can enhance their long-term timber production and environmental conservation programs by identifying and establishing a permanent forest estate that permits adequate conservation of all vertebrate species. Timber forests need to be managed in consultation with wildlife biologists. With few exceptions, resources made available for vertebrate conservation should be directed toward identification, acquisition, and management of unprotected key forest areas in preference to translocation or captive breeding.

REFERENCES

Ammann, H. 1985. Contributions to the ecology and sociology of the Javan rhinoceros (*Rhinoceros sondaicus* Desm.). Ph.D. diss., University of Basel, Switzerland.

Appanah, S. 1985. Flowering in South-East Asian rain forests. *J. Trop. Ecol.* 1: 225–240.

Beaman, R. S., J. H. Beaman, C. W. Marsh, and P. V. Woods. 1985. Drought and forest fires in Sabah in 1983. *Sabah. Soc. J.* 8: 10–30.

Becker, P., M. Leighton, and J. Payne. 1985. Why tropical squirrels carry seeds out of source crowns. *J. Trop. Ecol.* 1: 183–186.

Becker, P., and M. Wong. 1985. Seed dispersal, seed predation and juvenile mortality of Aglaia sp. (Meliaceae) in lowland dipterocarp rainforest. *Biotropica* 17: 230–237.

Bennett, E. L., and Z. Dahaban. This volume. Wildlife responses to disturbances in Sarawak and their implications for forest management.

Bennett, E. L., and A. C. Sebastian. 1988. Social organization and ecology of proboscis monkeys (*Nasalis larvatus*) in mixed coastal forest in Sarawak. *Int. J. Primatol.* 9: 233–255.

Caldecott, J., and M. Kavanagh. 1983. Can translocation help wild primates? *Oryx* 17: 135–139.

Chivers, D. J., editor. 1980. *Malayan Forest Primates: Ten Years' Study in Tropical Rain Forest.* Plenum Press, New York.

Cox, P. A., T. Elmquist, E. D. Pierson, and W. E. Rainey. 1991. Flying foxes as strong interactors in South Pacific island ecosystems: A conservation hypothesis. *Cons. Biol.* 5: 448–454.

Davies, A. G., and J. Payne. 1982. *A Faunal Survey of Sabah.* WWF Malaysia, Kuala Lumpur.

Duff, A. B., R. A. Hall, and C. W. Marsh. 1984. A survey of wildlife in and around a commercial tree plantation in Sabah. *Malay For.* 47: 193–213.

Emmons, L. H., N. Jamili, and B. Alim. 1991. The fruit and consumers of *Rafflesia keithii* (Rafflesiaceae). *Biotropica* 23: 197–199.

Foenander, E. C. 1952. *Big Game of Malaya.* Batchworth Press, London.

Fujita, M. S. 1988. Economic importance of bat/plant interactions in palaeotropic regions. WWF-US. Report.

Galdikas, B. M. F. 1982. Orang utans as seed dispersers at Tanjung Puting, Central Kalimantan: Implications for conservation. In *The Orang Utan: Its Biology and Conservation.* L. E. M. de Boer, editor. W. Junk, The Hague, The Netherlands.

Gould, E. 1978. Foraging behaviour of Malaysian nectar-feeding bats. *Biotropica* 10: 184–193.

Goulding, M. 1980. *The Fishes and the Forest.* University of California Press, Berkeley.

Hubback, T. R. 1939. The two-horned Asiatic rhinoceros. *J. Bombay Nat. Hist. Soc.* 40: 594–617.

Inger, R. F., and P. K. Chin. 1962. *The Fresh-Water Fishes of North Borneo.* Chicago Natural History Museum, Chicago.

Johns, A. D. 1989. *Timber, the Environment and Wildlife in Malaysian Rain Forests.* Institute of South-East Asian Biology, University of Aberdeen, Scotland.

Kemp, A. C., and M. I. Kemp. 1974. Report on a study of hornbills in Sarawak, with comments on their conservation. WWF Malaysia, Kuala Lumpur.

Lambert, F. 1989. Fig-eating by birds in a Malaysian lowland rain forest. *J. Trop. Ecol.* 5: 410–412.

———. 1990. Avifaunal changes following selective logging of a North Bornean rain forest. Report, University of Aberdeen, Scotland.

———. 1991. Fruit-eating by the purple-naped sunbird *Hypogramma hypogrammicum* in Borneo. *Ibis* 133: 425–426.

Leighton, M., and D. R. Leighton. 1983. Vertebrate responses to fruiting seasonality within a Bornean rain forest. In *Tropical Rain Forest Ecology and Management*. S. L. Sutton, T. C. Whitmore, and A. C. Chadwick, editors. Blackwell Scientific, Oxford.

Lieberman, M., and D. Lieberman. 1986. An experimental study of seed ingestion and germination in a plant-animal assemblage in Ghana. *J. Trop. Ecol.* 2: 113–26.

Mackie, C. 1984. The lessons behind East Kalimantan's forest fires. *Borneo Res. Bull.* 16, 2: 63–74.

Marsh, C. W., and W. L. Wilson. 1981. *A Survey of Primates in Peninsular Malaysian Forests*. Universiti Kebangsaan, Malaysia, and Cambridge University, Cambridge.

McNaughton, S. J. 1988. Mineral nutrition and spatial concentrations of African ungulates. *Nature* 334: 343–345.

Medway, Lord. 1983. *The Wild Mammals of Malaya (Peninsular Malaysia) and Singapore*. Second edition. Oxford University Press, Kuala Lumpur.

Mohsin, A. K. M., and M. A. Ambak. 1983. Freshwater fishes of Peninsular Malaysia. *Malaysian Appl. Biol.* 6: 75–78.

Mykura, H. F. 1989. Hydrology, geomorphology and erosion potential. In *Expedition to Maliau Basin, Sabah, April–May 1988*. C. W. Marsh, editor. Yayasan Sabah Forestry Division/WWF Malaysia, Kota Kinabalu, Sabah.

Payne, J. B. 1979. Synecology of Malayan tree squirrels with special reference to the genus *Ratufa*. Ph.D. diss., Cambridge University, Cambridge.

———. 1988. *Orang-utan Conservation in Sabah*. WWF Malaysia, Kuala Lumpur.

———. 1992. Rarity and extinctions of large mammals in Malaysian rainforests. In *In Harmony with Nature, Proceedings of the International Conference on Conservation of Trop Biodiversity, 12–16 June 1990*. S. K. Yap and S. W. Lee, editors. Malayan Nature Society, Kuala Lumpur.

Payne, J., C. M. Francis, and K. Phillipps. 1985. *A Field Guide to the Mammals of Borneo*. The Sabah Society with WWF Malaysia, Kota Kinabalu, Sabah.

Rabinowitz, A. R. 1989. The density and behaviour of large cats in a dry tropical forest mosaic in Huai Kha Khaeng wildlife sanctuary, Thailand. *Nat. Hist. Bull. Siam Soc.* 37: 235–251.

———. 1990. Fire, dry dipterocarp forest, and the carnivore community in Huai Kha Khaeng wildlife sanctuary, Thailand. *Nat. Hist. Bull. Siam Soc.* 38: 99–115.

Rabinowitz, A. R., and S. R. Walker. 1991. The carnivore community in a dry tropical forest mosaic in Huai Kha Kheng Wildlife Sanctuary, Thailand. *J. Trop. Ecol.* 7: 37–47.

Ridley, H. N. 1894. On the dispersal of seeds by mammals. *J. Straits Branch Roy. Asiatic Soc.* 25: 11–32.

———. 1930. *The Dispersal of Plants throughout the World*. Reeve, Ashford, England.

Sheldon, F. H. 1986. Habitat changes potentially affecting birdlife in Sabah, East Malaysia. *Ibis* 128: 174–175.

Short, J. 1981. Diet and feeding behaviour of the forest elephant. *Mammalia* 45: 177–185.

Smythies, B. E. 1981. *The Birds of Borneo*. Third edition. Sabah Society with Malayan Nature Society, Kota Kinabalu, Sabah.

Start, A. N., and A. G. Marshall. 1976. Nectivorous bats as pollinators of trees in Malaysia. In *Tropical Trees: Variation, Breeding and Conservation*. J. Burley and B. T. Styles, editors. Linn. Soc. Symp. Ser. 2.

Stuebing, R. B., and J. Gasis. 1989. A survey of small mammals within a Sabah tree plantation in Malaysia. *J. Trop. Ecol.* 5: 203–214.

Terborgh, J. 1986. Keystone plant resources in the tropical forest. In *Conservation Biology: The Science of Scarcity and Diversity*. M. E. Soulé, editor. Sinauer Associates, Sunderland, Mass.

Whitten, A. J., J. D. Sengli, A. Jazanul, and H. Nazaruddin. 1984. *The Ecology of Sumatra*. Gadjah Mada University Press, Indonesia.

5

WILDLIFE RESPONSES TO DISTURBANCES IN
      SARAWAK AND THEIR IMPLICATIONS FOR FOREST
      MANAGEMENT
      Elizabeth L. Bennett and Zainuddin Dahaban

Until about fifty years ago, large areas of Sarawak were relatively undisturbed. The timber industry was in its infancy and had not yet had much impact on Sarawak's forests (Bugo, this volume; WWF Malaysia 1985). Rural people had been practicing shifting cultivation for hundreds of years (Hatch 1982). Their technology was limited, so the effect of agriculture was mainly restricted to areas near major rivers. By the late 1980s, the picture was very different. Technological advances had allowed the timber industry to spread through much of interior Sarawak. By 1989, 60% of the land area had been licensed for selective timber extraction (Kavanagh, Rahim, and Hails 1989), and by 1990, 4.1 million ha of production forest had been logged (Mok, Jalil, and Jiwan 1991). Rural people had access to outboard motors and logging roads, which facilitated travel, and chainsaws, which enabled them to clear larger areas of forest for cultivation. Both rural and town people had shotguns; more than 60,000 guns were legally registered in the state, and hunting was almost universal (Caldecott 1988). Rural people hunted for food and, in some areas, to supply town markets; people in towns and logging camps hunted for food and for sport (Caldecott 1988). The human impact on Sarawak's forests and wildlife had thus increased greatly within a very short time.

Preliminary findings of several related studies of different duration indicate that animal responses to varied uses of habitat may affect the ability of both animal and plant species to recover from human activities. This chapter is intended not as a criticism of current practices but as an exploration of the effects of disturbance on wildlife populations. This investigation is essential if management programs designed to benefit both wildlife and humans are to be based on solid foundations.

General Surveys

All data were collected in steep lowland or hill mixed dipterocarp forest (see Whitmore 1984 for definition). Eight surveys were conducted in different sites throughout Sarawak. Two additional surveys were carried out in Brunei, where shotguns are illegal, rural populations are sparse, and hunting of canopy animals is negligible. Sites were selected to sample different types and degrees of disturbance and hunting pressure, as well as critical areas for management purposes (fig. 5.1).

Of the sites, eight were primary forest with minor disturbance to the habitat. Two further sites were old shifting-cultivation land, both inside Batang Ai National Park (fig. 5.2). Both had been cultivated some thirty years before, and both were contiguous with large areas of primary forest in Lanjak-Entimau Wildlife Sanctuary, in Sarawak (168,758 ha), and Gunung (Mt.) Behtuang dan Karimun Nature Reserve, in Kalimantan (about 300,000 ha).

Selective Logging Study

Data were collected in mixed dipterocarp production forest in Nanga Gaat (fig. 5.3). The main study area consisted of steep terrain between 250 m and 380 m elevation. The soils were red-yellow podzols. The forest was dominated by trees of the Dipterocarpaceae, especially *Shorea* species.

FIGURE 5.1 Map of Sarawak, showing the study sites.

Wildlife Responses to Disturbances in Sarawak

FIGURE 5.2 Mosaic of shifting-cultivation land and forest remnant patches outside the Batang Ai National Park area. Photograph by E. L. Bennett.

FIGURE 5.3 Hill dipterocarp forest immediately after selective logging at Nanga Gaat. Note that patches of forest alternate with large canopy gaps created by logging. Photograph by E. L. Bennett.

Data collection started in December 1989, when the main study area was primary forest, about 7 km from the nearest access road or disturbance. From December 1990 to March 1991, the area was selectively logged. Data collection continued until the end of November 1991, and a further survey was conducted in April 1992, a year after the felling had finished. Surveys were conducted in two other sites in the same concession area. One had been logged in December 1988 and was surveyed in April and November 1991 (two to three years later). The other had been logged in January 1987 and was surveyed in September and November 1991 (four to five years later). In all three areas, tractor logging was employed. Trees were cut with chainsaws, pulled by bulldozers to a loading site, loaded onto trucks, and transported by road to a log pond. Extraction rates depended on available volume. In the main study area, 54% of all trees at least 10 cm dbh (diameter at breast height) were removed or destroyed. This was likely to have been similar in all three sites. All three were part of the same forest block, the maximum distance between them was just 35 km, and the terrain and slope were similar in all. The flora and fauna before logging were probably similar in all three sites.

Before logging, the area was remote—about 40 km from the nearest human settlement. Hunting would have been light. Once access roads were built, and during the logging process, hunting with shotguns occurred (figs. 5.4

FIGURE 5.4 Rural hunter with freshly killed barking deer. Photograph by E. L. Bennett.

FIGURE 5.5 Wild bearded pig meat for sale at the Kapit market. Photograph by E. L. Bennett.

and 5.5). Two months after logging finished, hunting ceased in the main study area because a bridge along the access road collapsed. Even during the logging process, hunting in the main study area was lighter than elsewhere because the presence of one of the authors acted as a deterrent to hunters, who were concerned both about its legality and his safety. Any animal seen was usually shot at, including ungulates, primates, the Malayan sun bear (*Helarctos malayanus*), and the clouded leopard (*Neofelis nebulosa*). Night hunting with vehicles and spotlights occurred in any area with road access and was especially heavy in weeks preceding local festivals.

METHODS

Animal abundance was assessed via line transect or open-strip survey (Brockelman and Ali 1987; Burnham, Anderson, and Jaake 1980; Davies and Payne 1982; Marsh and Wilson 1981). Data were analyzed using Fourier series analysis (Burnham, Anderson, and Jaake 1980). One or two survey transects 2 km long were cut at each survey site. Existing trails were not selected because they might have been used by local hunters or disturbed in some other way. In the logging study, new trails were also cut. Old logging roads were not used for surveys because they give a biased sample of the forest and its wildlife. The open space and colonizing vegetation along roads are not representative of the forest

FIGURE 5.6 Wrinkled hornbills are frequently hunted for their meat or feathers, which are used in traditional costumes. Photograph by E. L. Bennett.

as a whole; some species are attracted to it, while others avoid it. Hunting pressure is also greater along roads.

Surveys began within half an hour of dawn. Average walking rate was 500 m/hr, and 1-minute stops were made every 25 m. A minimum of 16 km was walked at every site. In the logging study, surveys were conducted each month in the main study area. All animals that could be identified by sight or sound were noted. More detailed data (such as numbers and path-to-animal distances) were recorded on the target species: all mammals, pheasants (Phasianidae), and hornbills (Bucerotidae) (fig. 5.6). Night surveys for nocturnal mammals were also conducted at as many sites as possible. These were done for several nights at all sites except: (1) those where there were night hunters (Tanjung Datu), where surveying would have been dangerous; or (2) those where access to the survey trail was by a long, steep walk, where the observer was too exhausted to climb it both morning and night (Gunung Gading and Kuala Belalong).

To estimate the impact of hunting in different areas, sites were categorized by hunting pressure: none or light, medium, and high. Criteria used to assess hunting pressure were: (1) ease of access—a function of proximity to a main road and distance from a center of population by boat or foot; and (2) the number of hunting parties detected in the area during the survey. Four sites had none or light hunting pressure, four had medium pressure, and two had high pressure. Data were analyzed using the Mann-Whitney U Test and the Kruskal-Wallis one-way analysis of variance (Siegal 1956).

RESULTS

Shifting Cultivation

The number of species of diurnal primates, squirrels, all mammals, and all birds was slightly higher in two old cultivation areas than in primary forest (fig. 5.7), although the number of species of hornbills was slightly higher on average in primary forest. No differences were statistically significant. The same trend was apparent when comparing the two old shifting-cultivation sites at Batang Ai with an area of primary forest that was part of the same forest block, Sungai Beloh.

Average densities of primates were similar in primary and old shifting-cultivation forests (fig. 5.8). There was a slight shift in relative species abundance, with langurs (*Presbytis* spp.) and gibbons (*Hylobates muelleri*)

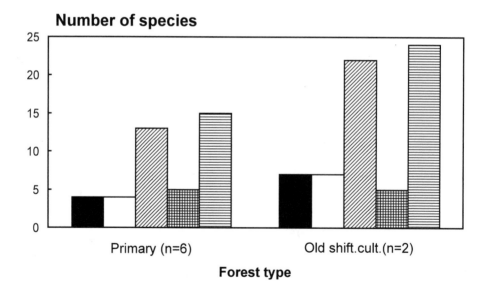

FIGURE 5.7 Number of species of different types of animals in primary forest and shifting-cultivation land regenerated for thirty years. Insectivores, bats, and murids (rats and mice) are excluded from the category "All mammals."

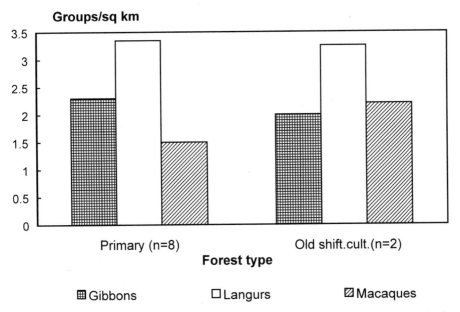

FIGURE 5.8 Density of primates (groups/km²) in primary forest and shifting-cultivation land regenerated for thirty years.

slightly more abundant in primary forest, and macaques (*Macaca* spp.) more abundant in old cultivation. Again, the differences were not statistically significant.

A more obvious difference in squirrel densities occurred between primary and old cultivation forests (fig. 5.9). The density of giant squirrels (*Ratufa affinis*) was higher in primary forest, though the difference was not significant. The density of medium-sized *Callosciurus* species was similar in both, but that of small *Sundasciurus* squirrels was significantly greater in old cultivation ($p < 0.05$).

Calling frequencies of large hornbills (helmeted [*Buceros vigil*], rhinoceros [*B. rhinoceros*], wrinkled [*Aceros corrugatus*], and wreathed [*A. undulatus*]) and medium-sized hornbills (Asian black [*Anthracoceros malayanus*], oriental pied [*A. coronatus*], white-crested [*Aceros comatus*], and bushy-crested [*Anorrhinus galeritus*]) were similar in primary forest and old shifting-cultivation sites (fig. 5.10). Argus pheasants (*Argusianus argus*), however, called significantly more frequently in old cultivation areas ($p < 0.05$). Increased calling might be due only to disturbance. In these areas, however, cultivation took place so long ago that disturbance was unlikely to have affected calling levels. High calling levels probably indicate high densities.

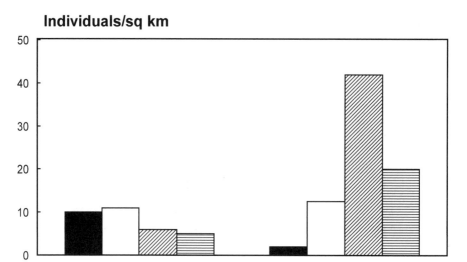

Individuals/sq km

Forest type

■ Ratufa   □ Callosciurus   ▨ Sundasciurus   ▤ Exilo/Nannosciurus

FIGURE 5.9 Density of squirrels (individuals/km²) in primary forest and shifting-cultivation land regenerated for thirty years.

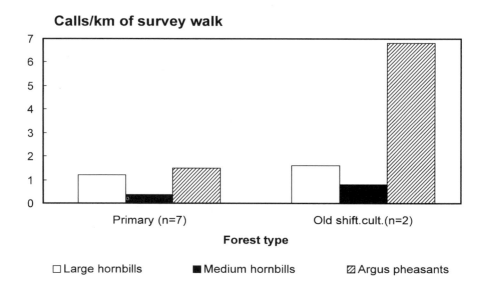

Calls/km of survey walk

Forest type

□ Large hornbills   ■ Medium hornbills   ▨ Argus pheasants

FIGURE 5.10 Frequency of calling (calls/km of survey walk) by large and medium-sized hornbills and by argus pheasants in primary forest and shifting-cultivation land regenerated for thirty years.

Logging

One year after logging, no significant change in the total number of animal species was detected (fig. 5.11). Species composition did change somewhat, however. For example, the four-striped ground squirrel (*Lariscus hosei*), rail babbler (*Eupetes macroceros*), and Burmese brown tortoise (*Geochelone emys*) appeared before logging but were not seen after. It is not possible to draw definite conclusions about the status of rarely seen rainforest animals, but they were probably extremely rare after cutting. Conversely, such secondary forest species as the magpie robin (*Copyschus saularis*), slender or lesser tree shrew (*Tupaia gracilis* or *T. minor*), and some bulbuls were never seen before logging but were recorded afterward. In the two- and four-year-old logged forest, diversity of primates and squirrels was similar to primary forest, although overall mammal diversity was lower (fig. 5.11).

Numbers of gibbons and red langurs (*Presbytis rubicunda*, also called maroon langurs) were lower one year after cutting (fig. 5.12). The density of white-fronted langurs (*Presbytis frontata*) also decreased, but not significantly.

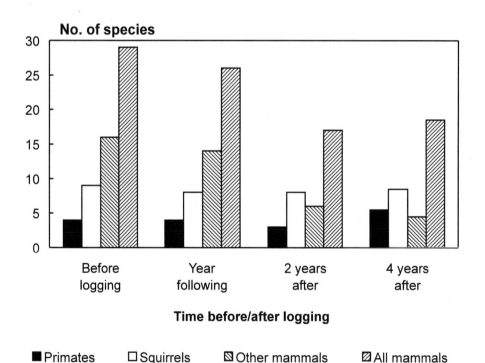

FIGURE 5.11 Number of species of different types of animals in relation to logging at Nanga Gaat. Species were observed in the main study area before logging and in the year after logging, and in other parts of the concession logged two and four years previously. Insectivores, bats, and murids are excluded from categories of "Other mammals" and "All mammals."

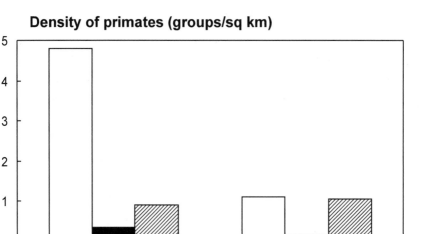

Density of primates (groups/sq km)

Time in relation to logging

☐ Bornean gibbon   ■ Red langur   ▨ White-fronted langur

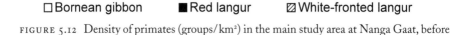

FIGURE 5.12 Density of primates (groups/km²) in the main study area at Nanga Gaat, before logging and in the year after logging.

Numbers of giant squirrels (*Ratufa affinis*) declined, whereas those of smaller squirrels (*Sundasciurus, Exilosciurus,* and *Nannosciurus)* increased (fig. 5.13). Ungulates declined initially, but by two years after logging numbers had increased above pre-logging levels. Densities of animals in the two- and four-year-old logged forest are not included because pre-logging densities are not known. Even though the forest and logging methods were similar, animal densities could vary naturally over the distances involved. A salt lick in the two-year-old logged forest, for example, would undoubtedly have influenced the abundance of certain species.

Hunting

With high levels of hunting, the number of species of primates and hornbills in an area declined, although the total number of mammal species present did not change significantly (fig. 5.14). The total number of bird species also declined. The effect of hunting on animal abundance was much more dramatic. For primates (fig. 5.15), numbers of gibbons and langurs decreased as hunting levels increased. Gibbons were absent in heavily hunted areas. Macaque densities were almost identical in areas with light and medium hunting, but none

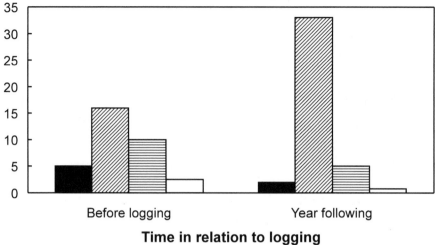

Density of mammals (indivs/sq km)

Time in relation to logging

■ Giant squirrel   ▨ Other squirrels   ☰ Tree shrews   ☐ Barking deer

FIGURE 5.13 Density of squirrels, tree shrews, and barking deer (individuals/km²) in the main study area at Nanga Gaat, before logging and in the year after logging.

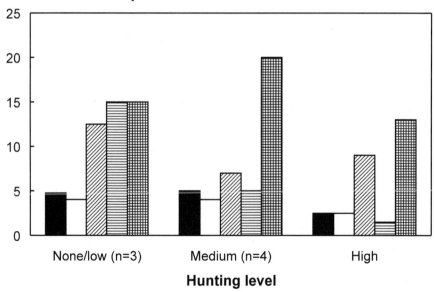

Number of species

Hunting level

■ Diurnal primates   ☐ Squirrels   ▨ All mammals   ☰ Hornbills   ▦ All birds

FIGURE 5.14 Number of species of different types of animals in forests with different levels of hunting intensity. Insectivores, bats, and murids are excluded from the category of "All mammals."

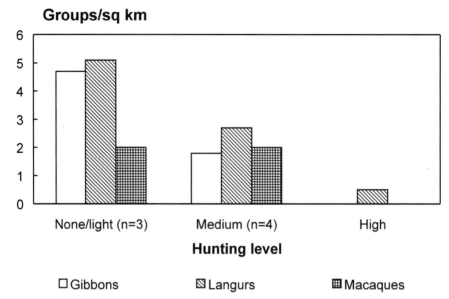

FIGURE 5.15 Density of primates (groups/km²) in forests with different levels of hunting intensity.

were recorded during formal surveys at heavily hunted sites. Because sites varied considerably, differences for any one species were not statistically significant. Densities of the total number of primate groups, however, were significantly different between sites of different hunting levels ($p < 0.05$). Moreover, differences in the number of individual primates were likely to be even greater than in the number of groups (numbers of individuals recorded were not accurate enough to use and would be biased against heavily hunted areas because of increased animal shyness). The number of individuals would be expected to decline sooner, and possibly more extremely, than the number of groups. Thus, the true effect of hunting is probably even greater than that shown here.

The effect of hunting on large squirrels was the opposite of its effect on primates (fig. 5.16). The number of giant and medium-sized squirrels (*Callosciurus* spp.) increased significantly with hunting pressure ($p < 0.05$). In areas where large mammals are abundant, squirrels are rarely hunted (Bennett, unpublished data; Caldecott 1988). The removal of many large fruit-and seed-eating animals (primates, hornbills, pigeons) would increase the food available for squirrels, so that their numbers might have risen owing to competitive release. Their densities could also have increased because of a reduction in the numbers of predators, particularly cats and civets. Both effects probably occurred simultaneously.

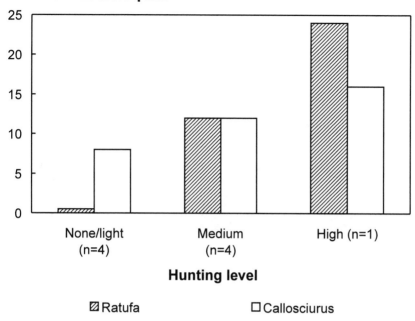

Individuals/sq km

FIGURE 5.16 Density of squirrels (individuals/km²) in forests with different levels of hunting intensity.

Hornbill numbers showed a small decrease in areas with medium levels of hunting but a great decline in areas with heavy hunting (fig. 5.17). Differences among sites of varied hunting intensity were not significant. Changes in one area could, however, be masked by different hunting patterns between areas. Town sport hunters and seminomadic Penan shot hornbills; feathers were found at campfires in Kubah National Park (Bennett and Walsh 1988) and in the Pulong Tau area (Bennett, unpublished data). These hunters presumably either ignored or were unaware of the animals' protected status. In other areas, such as Batang Ai National Park, people were more aware of the law, so probably did not hunt hornbills as intensively. Moreover, birds can easily fly into hunted areas, so the effects of hunting could be even greater than recorded.

The number of argus pheasant calls declined steadily with increasing hunting pressure, although variation among sites meant that differences were not significant (see fig. 5.17). Again, changes in one area could be masked by differences in hunting patterns between areas.

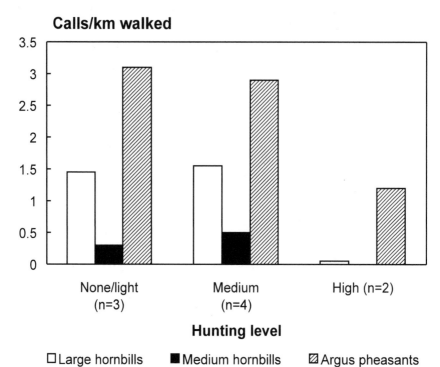

Calls/km walked

	None/light (n=3)	Medium (n=4)	High (n=2)

Hunting level

□ Large hornbills   ■ Medium hornbills   ▨ Argus pheasants

FIGURE 5.17 Frequency of calling (calls/km of survey walk) by large and medium-sized hornbills and by argus pheasants in forests with different levels of hunting intensity.

DISCUSSION

In previous studies of the effects of logging on mammals in Malaysian hill dipterocarp forests (Johns 1983, 1989, 1992; Johns and Skorupa 1987), wildlife survival was apparently greater than in the present study, even though timber extraction rates were similar (Johns and Marshall 1992). Gibbons, for example, fared far worse than previously reported. Part of the difference was probably methodological. First, previous studies compared results from different sites without accounting for possible differences in original densities between sites. Second, in all his studies in logged forests, Johns used logging roads as his survey routes. This method is likely to bias the results because some species might be attracted to roads (such as sambar deer, *Cervus unicolor*) and others avoid them (such as gibbons). In our study, random transects were cut to avoid such biases.

Other factors could also have contributed to different results. Original densities of wildlife were lower in the current study area than in Johns's study sites (Tekam, Peninsular Malaysia, and Ulu Segama, Sabah). Lower natural densities might mean that animals were more vulnerable to disturbance. Finally, levels of hunting might have been greater in our study. Both the observer's presence and the collapse of the access bridge after logging, however, meant that hunting was light in the main study area, and so was unlikely to have been the sole cause of differences from previous studies. Whatever the reason for the differences, the long-term effects of logging on wildlife clearly are not yet fully understood. Extreme caution is needed in interpreting existing data. For the species that either declined greatly in numbers or disappeared after cutting, moreover, it is not possible from the current data to say why they did. In the case of larger animals, it was impossible to distinguish between the effects of logging and hunting. It does seem, however, that such species are likely to need primary forest for survival and from which to recolonize disturbed areas.

In general terms, wildlife communities show similar responses to shifting cultivation and selective logging. In both types of deforestation, some species decline in number or become locally extinct, some remain stable, others increase in number, and a few new species enter the area. The trend with both types of disturbance is for edge or colonizer species to replace primary forest species. Thus, even though overall species diversity is similar in disturbed areas, the change is not necessarily beneficial from the standpoint of conservation. The species most threatened in Sarawak are those that depend on tall forests. Animals entering disturbed sites, such as the plantain squirrel (*Callosciurus notatus*), magpie robin (*Copyschus saularis*), and yellow-vented bulbul (*Pycnonotus goiavier*), are common in gardens throughout southeast Asia and unlikely to become extinct. As disturbance increases, they replace species that depend on undisturbed forest and cannot survive elsewhere. The aim of conservation is to maintain the total number of species not in a particular small area but within a country or region. It is well known that localized diversity can often be increased by disturbing the habitat to create more "edges" (Murphy 1989). This method, however, does not increase biodiversity in the whole of Sarawak; in the long run, widespread disturbance decreases diversity. Moreover, the two sample areas of old shifting-cultivation land were not typical of all such land throughout Sarawak. The most recent cultivation in these areas was more than thirty years ago, and both were contiguous with a block of about 300,000 ha of primary forest. There had thus been ample opportunity for animals to recolonize the disturbed areas, which is not commonly the case elsewhere in Sarawak. In addition, the two study areas were sufficiently close to primary forest that some species might still depend on primary forest for breeding. Orangutans (*Pongo pygmaeus*), for example, occur in disturbed habitats, but their breeding is suppressed in such areas (MacKinnon 1974); they might depend on access to primary forest for long-term survival (Bennett 1991). The results do indicate, however, that regenerated shifting-cultivation lands and logging areas can be important for wildlife conservation. The features of these areas that are important in allowing wildlife to persist are retention of some blocks of primary forest and control of hunting.

Hunting is a critical threat to wildlife in Sarawak. The numbers of all large mammals and birds studied were much lower in hunted areas than in unhunted ones. Hunting can be sustainable, even if the harvested population is kept below carrying capacity, provided that hunting does not exceed production (Caughley 1977). For certain ungulate populations, the maximum sustainable yield of a species occurs when hunting keeps the population at about 60% of carrying capacity (Caughley 1977). For slow-breeding species (such as some primates and hornbills), maximum yield is likely to be maintained when the population is far nearer to carrying capacity (Robinson and Redford 1991). In any case, populations of primates and hornbills were apparently far lower than 60% of potential carrying capacity even in areas where hunting pressure was medium, indicating that hunting was well above sustainable levels. In heavily hunted areas many species had become extinct or were on the brink of disappearing. Hunting has also had a major effect on Sarawak's large animals in the past. The Sumatran rhino (*Dicerorhinus sumatrensis*) and banteng (*Bos javanicus*) were virtually exterminated in Sarawak by hunting during the colonial era, whereas the tapir (*Tapirus indicus*) is believed to have become extinct in Borneo in historical times through hunting (Medway 1977).

Loss of large animals from a forest is not simply a conservation problem either in itself or for the sake of maintaining a protein supply for rural people. Large vertebrates are an integral part of a tropical forest; they stimulate diversity as pollinators, dispersers, and grazers (Payne, this volume; Redford 1992; Sunquist 1992). Without them, the forest is likely to suffer a loss of diversity. In Sarawak, loss of large animals could profoundly affect the forest, but the ramifications are as yet unpredictable. Data from this study indicate that the number of squirrels (predominantly seed predators) increases when the number of primates (predominantly seed dispersers) decreases. Hunting has thus already upset the natural balance in those areas. The long-term repercussions of this are unknown but could be substantial. Two other examples show that loss of vertebrates in Sarawak affects the forest. First, flying foxes (*Pteropus vampyrus*) are the only known pollinators of some forest trees (Payne, Francis, and Phillipps 1985), yet they are heavily hunted in Sarawak (Bennett, unpublished data; Caldecott 1988). Their demise could lead to the ultimate extinction of those trees (Payne, this volume). Second, civets (Rabinowitz 1991) and hornbills (Leighton 1982) are major seed dispersers. Some species of both are widespread in disturbed forests; overhunting them could severely hinder forest regeneration.

The resilience of tropical forests to disturbance is still largely unknown. The degree of recovery from shifting cultivation and logging indicates that the system can withstand fairly substantial degrees of change. In all cases where these have been studied, however, hunting levels have been light. If the animals are still present, many of the interrelationships of the forest can be maintained. In the absence of large animals, many of those functions will fail. Sustainability of forest use depends on maintaining the processes that keep the forest alive, and many of those are animal-dependent. In parts of Sarawak where hunting is heavy, many large animals are either extinct or likely to be ecologically extinct (see Redford 1992). The forests may appear healthy, with a rich tree flora and myriad orchids, ferns, and lianas. Without the animals, it is far from certain how long they can survive.

MANAGEMENT IMPLICATIONS AND CONCLUSIONS

Effective hunting controls are needed. These are difficult to enforce in any rainforest country, and Sarawak is no exception. Every effort should be made, however, to reduce overall levels, to direct hunting away from vulnerable species, and to concentrate hunting on species more able to withstand harvesting. Primates and hornbills both have long birth intervals, small litters, and a late age of first reproduction. Pigs are likely to be more resilient to hunting because their higher reproductive output means that they can more easily replace animals lost to hunters (see Bodmar, Fang, and Ibanez 1989). Controls are essential to ensure that no species become extinct through hunting, that there is a continued supply of meat to rural people, and that the animals continue to maintain forest diversity.

An effective system of Totally Protected Areas (TPAs) is also needed, because some species do not survive well in disturbed forests (Abang Morshidi and Gumal, this volume). In addition, large primary forest areas should be retained as wildlife reservoirs within regions of logging and shifting cultivation. Measures are needed to ensure that wildlife survives throughout the permanent forest estate, not just inside TPAs. This is because:

1. Some animals, such as large cats and raptors, occur at low densities and require large areas to protect viable populations. It is impractical

to expect there to be sufficient TPAs large enough to protect them fully. If the TPAs are part of a larger forest block in which they are protected from overhunting, most species should survive.

2. Without animals, the forest itself will suffer in the long term from a loss of diversity.
3. Animals are vital in helping forest regeneration.

Wildlife and TPAs should be considered not in isolation but as integrated parts of forest planning. Sarawak's forests should be viewed as an interconnected network, with TPAs and uncut areas within shifting-cultivation and logging zones strictly preserving species and habitats and protecting the environment. They are also reservoirs from which animals and plants can recolonize surrounding forests of different uses, increasing in intensity from extraction of minor forest products to subsistence hunting to commercial timber extraction and shifting cultivation.

ACKNOWLEDGMENTS

We thank former State Secretary Tan Sri Datuk Haji Bujang Nor for permission to conduct the studies. Members of the Sarawak Forest Department gave invaluable support, especially Datuk Leo Chai, Philip Ngau Jalong, Abang Haji Kassim bin Abang Morshidi, Ngui Siew Kong, Francis Gombek, Melvin Gumal, and Haji Osman, plus the numerous staff members who participated in surveys. Also of enormous help were Patrick Braganza, B. Ross Ibbotson, David Manja, and other staff members of Batang Balleh Forest Enterprises, Mohd. Nordin Hj. Hasan, Lt. Col. Sharkawi Hj. Hasbie, Mikaail Kavanagh, Mary Pearl, and Martha Schwartz. Many people were generous with their ideas and discussion, especially Richard Bodmer, Alan Rabinowitz, Rajanathan Rajaratnam, Kent Redford, John Robinson, and Thomas Struhsaker. The work was funded by NYZS/The Wildlife Conservation Society (formerly Wildlife Conservation International) and the World Wide Fund for Nature–Malaysia. Four anonymous reviewers gave extremely helpful comments on the manuscript. Finally, we thank Barney Chan and the Tropical Forest Foundation for inviting us to participate in the symposium.

REFERENCES

Abang Morshidi, A. H. K., and M. T. Gumal. This volume. The role of totally protected areas in preserving biological diversity in Sarawak.

Bennett, E. L. 1991. Diurnal primates. In *The State of Nature Conservation in Malaysia*. R. Kiew, editor. Malayan Nature Society, Kuala Lumpur.

Bodmar, R. E., T. G. Fang, and L. M. Ibanez. 1989. Primates and ungulates: A comparison of susceptibility to hunting. *Primate Cons.* 9: 79–83.

Brockelman, W. Y., and R. Ali. 1987. Methods of surveying and sampling forest primate populations. In *Primate Conservation in the Tropical Rain Forest*. C. W. Marsh and R. A. Mittermeier,
editors. Alan R. Liss, New York.

Bugo, H. This volume. The significance of the timber industry in the economic and social development of Sarawak.

Burnham, K. P., D. R. Anderson, and J. L. Laake. 1980. Estimation of density from line transect
sampling of biological populations. *Wildl. Monogr. 72.*

Caldecott, J. O. 1988. *Hunting and Wildlife Management in Sarawak*. IUCN, Gland, Switzerland.

Caughley, G. 1977. *Analysis of Vertebrate Populations*. Wiley, Chichester, England.

Davies, A. G., and J. B. Payne. 1982. *A Faunal Survey of Sabah*. WWF Malaysia, Kuala Lumpur.

Hatch, T. 1982. *Shifting Cultivation in Sarawak—A Review*. Soils Division (Research Branch),
Department of Agriculture, Sarawak, Kuching.

Johns, A. D. 1982. Ecological effects of selective logging in a West Malaysian rain-forest. Ph.D.
diss., University of Cambridge.

———. 1989. Timber, the environment and wildlife in Malaysian rain forests. Final report to the
Institute of South-east Asian Biology, University of Aberdeen, Scotland.

———. 1992. Species conservation in managed tropical forests. In *Tropical Deforestation and
Species Extinction*. T. C. Whitmore and J. A. Sayer, editors. Chapman and Hall, London.

Johns, A. D., and A. G. Marshall. 1992. Wildlife population parameters as indicators of the
sustainability of timber logging operations. In *Forest Biology and Conservation in Borneo*.
Ghazally Ismail, Murtedza Mohamed, and Siraj Omar, editors. Yayasan Sabah, Kota
Kinabalu, Sabah.

Johns, A. D., and J. P. Skorupa. 1987. Responses of rain forest primates to habitat disturbance—A
review. *Intl. J. Primatol.* 8: 157–191.

Kavanagh, M., A. A. Rahim, and C. J. Hails. 1989. *Rainforest Conservation in Sarawak: An
International Policy for WWF*. WWF Malaysia, Kuala Lumpur, and WWF International, Gland,
Switzerland.

Leighton, M. 1982. Fruit resources and patterns of feeding, spacing and grouping among sympatric Bornean hornbills (Bucerotidinae). Ph.D. diss., University of California, Davis.

MacKinnon, J. R. 1974. The behavior and ecology of wild orang-utans (*Pongo pygmaeus*). *Anim.
Behav.* 22: 3–74.

Marsh, C. W., and W. L. Wilson. 1981. *A Survey of Primates in Peninsular Malaysian Forests*.
Universiti Kebangsaan Malaysia, Kuala Lumpur.

Medway, Lord. 1977. Mammals of Borneo. *Monogr. Malaysian Branch Roy. Asiatic Soc.* 7: 1–172.

Mok, S. T., A. Abdul Jalil, and D. Jiwan. 1991. *A WWF Strategy for Tropical Forest in Sarawak*. WWF
Malaysia, Kuala Lumpur, and WWF International, Gland, Switzerland.

Murphy, D. D. 1989. Conservation and confusion: wrong species, wrong scale, wrong conclusions. *Cons. Biol.* 3: 82–84.

Payne, J. This volume. Links between vertebrates and the conservation of Southeast Asian rainforests.

Payne, J., C. M. Francis, and K. Phillipps. 1985. *A Field Guide to the Mammals of Borneo*. Sabah
Society with WWF Malaysia, Kota Kinabalu and Kuala Lumpur.

Rabinowitz, A. 1991. Behavior and movements of sympatric civet species in Huai Kha Kaeng
Wildlife Sanctuary, Thailand. *J. Zool.* London 223.

Redford, K. H. 1992. The empty forest. *BioSci.* 42: 412–422.

Robinson, J. G., and K. H. Redford. 1991. Sustainable harvest of neotropical forest mammals. In
Neotropical Wildlife Use and Conservation. J. G. Robinson and K. H. Redford, editors.
University of Chicago Press, Chicago.

Siegal, S. 1956. *Non-parametric Statistics for the Behavioral Sciences*. McGraw-Hill Kogakusha,
Tokyo.

Sunquist, F. 1992. Blessed are the fruit eaters. *Intl. Wildl.* (May–June): 4–10.

Whitmore, T. C. 1984. *Tropical Rain Forests of the Far East.* Second edition. Clarendon Press, Oxford.

WWF Malaysia. 1985. *Proposals for a Conservation Strategy for Sarawak.* Compiled by L. Chan, M. Kavanagh, Earl of Cranbrook, J. Langub, and D. R. Wells. WWF Malaysia, Kuala Lumpur, and State Planning Unit of Sarawak, Kuching.

PLATE I   The Penan are a nomadic people dependent on the tropical forest. Many of them are now settled in villages, but they still use some forest products to supplement subsistence farming. Photograph by R. Primack.

The Sarawak Timber Association and Caterpillar, Asia, graciously provided support for including the color illustrations.

PLATE 2A   Dipterocarp species, the dominant timber trees of Southeast Asian rainforests, have characteristic winged fruits. Photograph by the Sarawak Forest Department.

PLATE 2B   Durians are a fruit tree characteristic of the region. Wild fruits are used as food by many animals, and cultivated fruits are important in the local market. Photograph by the Sarawak Forest Department.

PLATE 3A   Logging practices in dipterocarp forests rely on heavy machinery. Photograph by the Sarawak Forest Department.

PLATE 3B   Logging, a major source of employment in rural areas, is particularly labor-intensive in peatswamp forests because heavy machinery cannot be used on soft soils. Photograph by John Proctor.

PLATE 4 Poorly planned logging roads and felling on steep slopes can cause extensive soil erosion, delaying regrowth and damaging watersheds. Photograph by R. Primack.

PLATE 5   Asian pitcher plants in the genus *Nepenthes* grow in abundance in protected areas, such as Mt. Kinabalu National Park in Sabah, Malaysia. Elsewhere their populations have often been damaged and overcollected. Photograph by the Sarawak Forest Department.

PLATE 6A   Spectacular limestone pinnacles are included in Mulu National Park, Sarawak. Mulu National Park has a rich flora and fauna and is also being developed for ecotourism. Photograph by John Proctor.

PLATE 6B   Mt. Kinabalu is a center of biological diversity. Shown here is a giant moss, *Dawsonia superba*, in lower montane forest. Photograph by Wong Khoon Meng.

PLATE 7A   Dipterocarp seedlings often occur at high densities on the forest floor, growing rapidly when a tree falls and opens the canopy above them. Photograph by Pamela Hall.

PLATE 7B   Tree seedlings are grown in nurseries and then used for enrichment plantings in logged forests and for the reforestation of degraded lands. Photograph by R. Primack.

PLATE 8A   Insects, such as these trilobite beetle larvae, constitute the majority of species in tropical forests. Photograph by Wong Khoon Meng.

PLATE 8B   Tropical forests supply a wide range of nontimber products. Birds are collected using resin-covered sticks and then are sold as pets. Photograph by R. Primack.

6 RAINFORESTS AND THEIR SOILS
John Proctor

Rainforests and their soils are still often linked by the widely accepted ideas that the soils are invariably nutrient-poor, physically fragile, or both and that heavily exploited forest is likely to be slow to recover because its soil will not sustain rapid regrowth. These ideas, and several others relating to rainforests and their soils, are based on slender evidence (Proctor 1983, 1987, 1992a). They fit in well with conventional wisdom on rainforest conservation but are rarely critically evaluated because hard facts about the relations between rainforests and their soils are difficult to obtain.

Using examples from Southeast Asia as much as possible, in this chapter I examine some assumptions about rainforests and their soils and assess what we need to know about rainforest soils for successful forest management. More experimental work is needed, and I suggest a possible way forward in our understanding by looking at rainforests on limestones and ultramafic rocks. (Ultramafic rocks, often termed "ultrabasic" or "serpentine," are rich in magnesium and iron, poor in most plant nutrients, and have substantial quantities of nickel, which is potentially toxic for plants.)

NUTRIENTS IN BIOMASS AND SOILS

Quantifying rainforest nutrients poses formidable problems (Proctor 1983, 1987). First, adequate estimates of forest biomass are rare and usually relate only to above-ground parts. The biomass of large roots remains largely unknown, although good measurements of fine (less than 0.5 cm) roots are beginning to be made in the forest at the Danum Valley in Sabah (Green 1993). Second, there are large differences in nutrient composition between species (Grubb and Edwards 1982). Most nutrients in the biomass are in the relatively few large trees, which are likely to be locally rare and difficult to sample. It can take a week simply to dismember one large tree and prepare it for the chemical analysis of its constituent above-ground parts. Third, the subsequent chemical analysis is a lengthy procedure, and even well-funded studies often analyze only a few large trees.

Even the simpler problems of soil analyses have not yet been adequately worked out. Rainforests occur on a range of soils, some nutrient-rich, others poor in at least one nutrient. Nutrient-poor soils predominate, but there is no agreement on the proportion of total soil nutrients available to trees and often no attempt in studies to relate the depth of soil samples with those of root systems. It is gradually becoming evident that rainforest tree roots may obtain water and nutrients from well below the depth of conventional soil analyses. Nepstad (personal communication, 1992) has found functional roots down to 18 m in Brazil. Tropical soil surveys are likely to remain incomplete and widely accepted methods of soil classification are often of limited use for forest ecologists and managers (see Proctor 1992a).

DO RAINFORESTS CYCLE NUTRIENTS EFFICIENTLY?

The measurement of the flow of nutrients into and out of tropical forests poses many methodological difficulties. A finite quantity of nutrients enters the forest in rainwater, but even this is fairly difficult to measure accurately. The two methods of measuring outputs from the system are to analyze the movements and nutrient concentrations in soil water (lysimetry) or to measure the nutrient concentrations in streams flowing from the forest (the watershed approach). Both methods are beset with problems. Bruijnzeel (1991) provides a good review of data on tropical forest nutrient input and output and shows some of the conflicting results.

When interpreting input and output data one must consider that there are two fundamentally different types of rainforest. Some rainforests are on relatively shallow soils and have a contribution from rock weathering to their nutrient input; others are on soils that have bedrock beyond the rooting zone (Baillie 1989; Burnham 1989). Those on shallow soils are "open" systems with a propensity for net nutrient loss, whereas those on deep soils are "closed" systems in which nutrients are likely to cycle more tightly. In practice many rainforests lie somewhere on a cline between "open" and "closed." In Southeast Asia the shallow soils over shale, for example, may be intermediate and have features of both systems. Moreover, deep rooting might blur the distinction between the two systems. It is clear, however, that nutrient losses from the forest at large and losses from the biologically active part of the system must be distinguished. A watershed approach on a closed system might measure nutrient losses provided by the weathering bedrock and uncoupled from the forest above; for any particular forest, it may be impossible to quantify the efficiency with which it cycles nutrients.

Soils exert a dramatic influence on rainforests. Heath forests occur on acidic sandy soils and are variously known as bana, caatinga, campina, campinarana, and kerangas. Mangrove forests occur on tidally inundated saline soils, peatswamp forests occur locally where the drainage is poor, and stunted forests occur on some ultramafic soils. These variations bear testimony to the importance of soil types (fig. 6.1). Even when soils are less distinctive than those just mentioned, they may cause distinct differences in species composition. For example, Ho, Newbery, and Poore (1987), working in Peninsular Malaysia, showed large floristic differences between forest on the Segamat soil series and forest on the Batu Anam soil series about 3 km distant. They attributed these differences to the lower clay and nutrient concentrations and the greater water-holding capacity of the Batu Anam series.

Sometimes interactions between soils, climate, and other factors determine gross differences between vegetation types. Ash (1988) has shown that rainforest in northeastern Queensland is usually associated with soil with a high mineral status. Rainforest extends into drier regions on a basaltic (base-rich) substrate. However, Ash stresses that topography, climate, and burning patterns interact with soils in determining the presence of the rainforest. In northern Brazil, Thompson, Proctor, and Scott (1994) found that higher soil nutrients near the forest-savanna boundary on Maracá Island are associated with semi-evergreen rather than evergreen forest. Furley, Ratter, and Gifford (1988) and Ratter et al. (1973, 1978) have made similar observations linking greater deciduousness with higher soil nutrients elsewhere in Brazil.

The influence of soil factors has proved difficult to elucidate in many cases. Wong and Whitmore (1970), for example, could find no evidence of a link between forest composition and soil types in a 500-ha block of lowland dipterocarp forest at Pasoh in Peninsular Malaysia. Ashton (1976) reinvestigated the area of Wong and Whitmore (1970) using a greatly expanded sample size. He concluded that "floristic variation is unequivocally and consistently correlated principally with environmental factors, among which physiography is clearly important." Unfortunately, details of the soils were lacking; his classes of forest were associated with coarsely different classes of land: "hillside, alluvium, and lower hillside/undulating land." Further, Ashton (1976) agreed with Wong and Whitmore that many species do occur in groves more or less independently of variation in the physical environment. Using a range of multivariate statistical

FIGURE 6.1   Stunted forests on extreme soil types. *Top:* Forest on shallow soils over ultramafic rock. Photograph by J. Proctor. *Bottom:* Extreme stunted forest on shallow sandy soils at Bako National Park in Sarawak. Photograph by R. Primack.

techniques, Newbery and Proctor (1984) found that soil factors had a limited impact on the floristic composition within four 1-ha plots in Gunung Mulu National Park, Sarawak, although soil differences among the plots were the likely cause of their contrasting vegetation (Proctor et al. 1983). Studies in Brunei and Sarawak have suggested that floristic variation within dipterocarp-rich evergreen lowland rainforest varies predictably in relation to soil nutrients and topography (Austin, Ashton, and Greig-Smith 1972; Baillie and Ashton 1983; Baillie et al. 1987). Their results, however, are complex and hard to interpret. Consider, for example, the conclusions of Baillie et al. (1987): "When individual plots are treated as separate cases there are marked associations between the distribution of many species and site conditions. In the statistically more conservative results based on clusters of plots, the results appear to show little overall influence of site conditions on tree species distribution." In considering the different conclusions from the plots and clusters, they noted that "the choice of which set of results is taken as being closer to reality is subjective."

Where small-scale soil heterogeneity has been shown to be an important forest determinant, differences in topography or drainage have often been involved. An extreme example of the effect of topography was demonstrated across a forest-savanna boundary in Brazil by Thompson et al. (1992b). Though soil nutrient composition varied slightly from forest to savanna, the savanna boundary coincided with a slight depression where the wet-season water table came to the soil surface. Ho, Newbery, and Poore (1987) found vegetation classes within their 11.7-ha plot at Jengka, Malaysia, associated with relatively wet and relatively dry soils. Further evidence of the importance of topographic and drainage effects is found on small isolated mountains such as Mt. Silam in Sabah (Proctor et al. 1988). This mountain is just 880 m high, yet its forests range from dipterocarp-rich evergreen lowland rainforest at 280 m to stunted montane forest without dipterocarps at 870 m. Few species have a wide altitudinal range. The substantial vegetation changes up to the cloud cap at about 650 m seem to be determined by relatively slight changes in soil chemistry, soil water supply, or both. Above that altitude, the changes in both soils and vegetation are much greater and caused by a cloud cap (Bruijnzeel et al. 1993). These examples highlight the need for studies of site hydrology in any attempts to understand the influence of soil on vegetation.

The resolution of conflicting conclusions about how much soils account for differences in forests seems straightforward but uninformative. First, the greater the soil differences, the more likely there are to be differences in vegetation. Newbery and Proctor (1984), working on plots in Gunung Mulu

National Park, Sarawak, claim that "soil (chemical) differences needed to be in the order of three to five fold to have a detectable shift in floristic composition over a distance of about 100 m." Second, species are likely to vary in their responses to soil changes. Third, interactions between soils and other factors are likely to be important. Any attempt to correlate small-scale soil heterogeneity with floristic composition will admittedly always be confounded by the ability of plants to alter soils, so that correlations may not indicate causality. Also, because tree roots are likely to cover at least the area of the crown, they will integrate soils over a wide area.

SOILS, FOREST STATURE, AND PRODUCTION

Although knowledge of soil nutrient supply is imperfect, the available data suggest that little correlation exists between soil chemistry and forest stature. In Mulu, Sarawak, Proctor et al. (1983) found huge dipterocarp forests on nutrient-poor soils. Vitousek and Sanford (1987) have concluded from their review of rainforest mineral nutrition that "an association between soil fertility and above-ground biomass is . . . unlikely in any but the most extreme cases." Anderson and Spencer (1991) have commented that "there appears to be little evidence that the stature and productivity of mature forests are related to the inherent fertility of parent soil."

The view that soil nutrient supply limits forest production, if not biomass (Jordan 1985), is still widely accepted despite evidence to the contrary. The prevailing model suggests that rainforests in low-nutrient soils have low leaf nutrient concentrations, particularly low leaf-litterfall nutrient concentrations, and that the forests are unproductive. When Thompson et al. (1992b) tested the model in rainforest on nutrient-poor sandy soils on Maracá Island, Roraima, Brazil, it was contradicted in every way. The forest had large, mesophyllous leaves with relatively high nutrient concentrations, contrary to predictions. It also had a high litterfall production, and the litterfall, moreover, had relatively high nutrient concentrations. A comparison of soil and litterfall data from Maracá with those for La Selva, Costa Rica, showed that although the soil at La Selva had a much higher nutrient status, the forest produced roughly as much litterfall with broadly similar nutrient concentrations as that on Maracá. The extent of nitrogen fixation by leguminous trees was not known, but these species are unlikely to significantly contribute to the high nitrogen concentrations at Maracá because they accounted for only 3.2% of the basal area (trees at least 10 cm dbh). At La Selva, leguminous trees made up 33.9% of the basal area. Although rejecting a model when a single exception is found has dangers, the

Maracá example highlights the gulf between fact and understanding in rainforest mineral nutrition.

A further feature of the work at Maracá Island was the large number of foliar analyses conducted (Thompson et al. 1992b). These analyses showed not only a generally relatively high concentration of nutrients but also wide interspecific variations in nutrient concentrations. We still do not know why species should be so variable in this respect (similar interspecific variations have been shown in a montane forest in New Guinea by Grubb and Edwards 1982, mentioned above) and must admit that rainforest foliar analyses cannot be used as an indicator of ecosystem nutrient status. If one nutrient was of overriding limiting importance at Maracá, one might perhaps expect more interspecific uniformity, and uniformly low litterfall concentrations of that element. Neither case applies.

Some information about the relation of wood production by dipterocarps to soil nutrients has been given by Primack et al. (1987), who describe fifteen years of growth studies of trees in three primary forest sites at Bako, Lambir, and Bukit Mersing, all in Sarawak. Soils analyses showed a gradient of increasing nutrient concentrations from Bako to Lambir to Bukit Mersing. The gradient was not paralleled by the mean diameter increments, which were ranked for all dipterocarp species combined at each site: Lambir was less than Bako, which was less than Bukit Mersing. The authors concluded that "the anticipated relationship between soil fertility and growth is not demonstrable, and other factors such as local weather patterns, elevation, tree competition, pests and pathogens also play a role in controlling tree growth rates." Ashton and Hall (1992) conclude from their work at Bako, Lambir, and Bukit Mersing that the growth of younger trees, though not older ones, is correlated with soil nutrients. This conclusion does not seem justified, however, since much larger canopy gaps occurred at the nutrient-rich site, Bukit Mersing, and the more rapid growth of the young trees might be just as easily be correlated with higher light intensities as with higher soil nutrients.

MYCORRHIZAS

Widespread assumptions about the importance of mycorrhizal associations in tropical forests should be treated with caution. Although relations between improved plant growth and mycorrhizal infection have been obtained under field conditions in the tropics, these have involved crop species in plantations or pot experiments (Janos 1987; Santoso 1988). There are no data on the importance of mycorrhizas in the functioning of natural tropical forests. Fitter (1986) has called for field studies to examine the influence of mycorrhizal infections on survivorship, fecundity, and overall fitness of plant populations. But as Alexander (1989) has commented, "[In] tropical forests such studies seem a long way off and for some time to come we will have to continue to infer dependence of a given species from the morphology of its root system (Baylis 1975; Janos 1987), rather than from objective evidence." Although a number of successful applications of mycorrhizal (of the vesicular-arbuscular type) inoculations in temperate horticulture have been documented (Hayman 1987; Nemec 1987), these are largely confined to specialized systems in which soil sterilization or micropropagation are standard. After collecting data from a large number of field trials in which both changes in plant yield and in infection density had been reported, McGonigle (1988) could find no relation between the two variables. Fitter (1991) has discussed the difficulty of demonstrating that mycorrhizal infection benefits plants under natural conditions, and his views are reinforced when the complexity of tropical rainforests is considered.

Widespread assumptions about the importance of mycorrhizas in tropical forests are not justified on the present evidence. Mycorrhizas may be important, but much further work incorporating the suggestions of Fitter (1991) is needed before their role is proven.

FERTILIZATION AND BIOASSAY EXPERIMENTS

Grubb (1989) has drawn attention to the dearth of experimental work in the field of rainforest mineral nutrition. The only published fertilization experiments in natural rainforests have all involved montane forests: Gerrish, Mueller-Dombois, and Bridges (1988) and Vitousek et al. (1987) in Hawaii; Tanner et al. (1990) in Jamaica; and Tanner, Kapos, and Franco (1992) in Venezuela. The results show that nitrogen supply limits girth increments of many trees in these forests. The evidence for limitation by phosphorus supply is less strong and best for the site in Jamaica. Much scope for similar fertilization experiments in lowland forests remains, and Mirmanto, Green, and Proctor (unpublished data) have initiated a study at Barito Ulu in Central Kalimantan, Indonesia.

Another approach is to investigate tropical soils by bioassay. This method is likely to be most successful when native species are used. Denslow, Vitousek, and Schultz (1987) grew six species of shrub and one large herb in intact soil cores from the forest at La Selva, Costa Rica. A soil chemical analysis showed the soil to have a medium phosphate status. The experiments were conducted in a humid shade-house and included +N, +P, and complete nutrient fertilization. The shrubs were three species of *Miconia* (Melastomataceae) and three species of *Piper* (Piperaceae) and consisted of some that normally grow under deep shade and others that normally grow in high light conditions. The herb was *Phytolacca riviniana* (Phytolaccaceae), which is fast-growing and typical of open cultivated areas. No species responded to the addition of nitrogen, and phosphorus addition significantly increased the growth only of *Phytolacca*. But the addition of a complete fertilizer treatment increased growth in all cases, suggesting that elements other than nitrogen and phosphorus or some nutrient interaction might be limiting. In Southeast Asia, Turner (1991) has shown that fertilization by major nutrients resulted in substantially increased growth of the pioneer species *Melastoma malabathricum* (Melastomataceae) and *Trema tomentosa* (Ulmaceae). *Trema tomentosa* showed the greater response, which was associated with its usual occurrence on more nutrient-rich soils. By contrast, similar fertilization of four dipterocarp species did not increase growth (Turner, Brown, and Newton 1993).

The case for additional experiments is strong, but it is axiomatic that such experiments must be rigorous to permit the identification of individual effects and interactions. Fundamental questions also exist about the more general ecosystem effects of fertilization and how far the data on such species at a certain stage of their lives are generally applicable. The experiments are likely to give results that are difficult to interpret. The clear differences in species responses to added nutrients show the complexities of the problem, but experiments remain a promising approach to understanding the extent of limitation of rainforest growth by mineral nutrients.

NUTRIENTS AND LOGGING

The data on nutrient depletion by logging activity, though few and subject to the concerns about soil and tree analyses expressed earlier, at least suggest that moderate or even heavy logging will not result in nutrient limitation as long as factors contributing to soil erosion are limited. When well-established commonsense precautions (Bruijnzeel 1992) are taken against soil erosion there is no evidence that logging activities cause a catastrophic decline in soil nutrients. In an Australian rainforest Gillman et al. (1985) found that although logging decreased organic carbon by 15%, the quantities of total nitrogen did not change. Boxman et al. (1985) showed that careful logging caused no nutrient losses important enough to affect forest production. As Anderson and Spencer (1991) have commented, "Provided that extreme forms of mechanical disturbance during logging and site preparation for plantations are avoided, there is little evidence that nutrient depletion limits the rate of forest growth." Unfortunately, in many cases, extreme disturbance of soil followed by erosion and nutrient loss does often occur following logging and the establishment of plantations.

One of the best studies on nutrients and logging is that of Jonkers (1989), who showed in Suriname that for a timber yield of 23 m^3 ha^{-1} with a mass of 17 t ha^{-1} (equivalent to felling 32.6 t ha^{-1} stemwood, 15.3 t ha^{-1} leaves and branches, and leaving behind 30 t ha^{-1} as stumps, top ends, and defective wood), erosion losses were negligible and there was a harvest loss of about 50 kg ha^{-1} of nitrogen and 160 kg ha^{-1} of other nutrients in the logs. These quantities were very low compared with the estimated nutrient content of the living phytomass of virgin forest (about 2 t ha^{-1} nitrogen, 5 t ha^{-1} other nutrients). Jonkers felt that the small quantities removed by logging could be made up by rainfall, nitrogen fixation, and uptake from the slowly accreting soil pool. Jonkers considered that "harvesting 46 m^3 ha^{-1} (a very high value for Suriname) results in a doubling of the loss of nutrients, but it is believed that the ecosystem can cope also with this loss although full recovery may take longer than the felling cycle of 20 years." Poels (1987) has shown that stream water losses of nutrients from logged-over forest in Suriname were slight.

It is important to realize that the analysis of felled trees may not give a guide to limiting nutrients and the trees are likely to have more nutrients in them than they actually require. Studies of nutrient removal by logging activities will hence tend to overestimate the impact of logging on forest production. Ideally such studies should always be combined with an assessment of the actual nutrient requirements of trees, but this is a difficult task. The best approach to studying the importance of mineral nutrients for recovery after logging is probably an experimental one with additions of nutrients in various combinations in carefully designed experiments on forest that has been logged at differing intensities.

NUTRIENTS IN PLANTATION FORESTRY

Evans (1992) has summarized much information about the use of fertilizers in plantations with rapid tree growth and high nutrient demands. With short rotation crops on poor soils, nutrient depletion is a real possibility, especially in the moist tropics. Rotations are short, which makes fertilizing more economical. Increasingly just one or two species are used in afforestation, for easier management and a more uniform end product. These species may not be suited to all sites encountered, and some nutrient input is often necessary to aid their establishment. On some impoverished sites, the addition of small amounts of nutrients have produced spectacular improvements in growth, such as boron for eucalypts growing in many African grasslands and zinc for pines in Madagascar and northern Australia (Rampanana et al. 1988; Rance, Cameron, and Williams 1982).

Studies on two neglected forest formations—forest over limestone and forest over ultrabasic (ultramafic) rock (Whitmore 1984)—may enhance our knowledge of rainforest mineral nutrition. Although these formations cover a relatively small total area, they occur in many tropical areas and are scattered throughout the tropical Far East (fig. 6.2). They have a limited but not negligible interest for commercial exploitation and are of outstanding conservation interest because of their abundance of endemic species and specialized local races. They offer ideal subjects for experimental work because many of them are so unusual chemically. Just as much has been learned, for example, about plant-soil relations in temperate areas (such as Kruckeberg 1992) from comparative studies of plants of these rocks and of those of contrasting soils, so there is great potential for similar studies on tropical examples. These tropical formations are likely to be unrecognized reservoirs of genetic diversity; they are taxonomically under-investigated, and extrapolation from temperate areas (such as Proctor 1971) suggests widespread inherited adaptations that may or may not be expressed morphologically. Several of these tropical rock formations have some of the world's most remarkable vegetation and plant species—and some of its most remarkable soils. One example is the shallow, highly organic (95% loss-on-ignition) soils of the Gunung Api limestone, Sarawak, which have a near neutral pH (Proctor et al. 1983). These soils seem to exist because organic

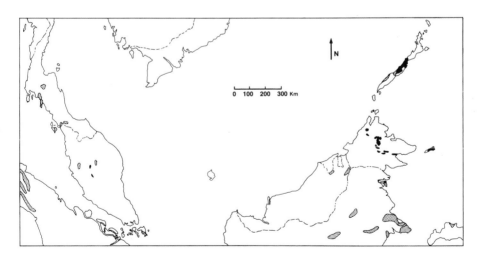

FIGURE 6.2   This map shows the extent of forest formations over limestone (dotted) and ultramafic (shaded) bedrock in Malaysia and parts of Indonesia and the Philippines. These formations, sometimes covering much larger areas than the ones depicted, are found in many other parts of Southeast Asia.

Rainforests and Their Soils   97

decomposition is limited by excessive drainage and droughtiness due to a lack of mineral soil from the very pure limestone.

Little remains known about forests over limestone in the Far East, and they seem to have been neglected in recent years. Crowther (1982) has made some valuable ecological observations in West Malaysia, but the situation described by Chin (1977), who has produced the most detailed account of the vegetation in this region, still applies: "In relation to its rich flora, extremely varied habitat and the fact of uneven local exploration, the limestone vegetation is probably the least botanized, and hence botanically known, of all the vegetation types in Malaya." The same applies to the geologically different limestones of Sarawak and those elsewhere in the region. Among the few accounts are Anderson (1965), the descriptions of the forest on Gunung Api described by Collins, Holloway, and Proctor (1984) and Proctor et al. (1983), and plots on soils with a varying limestone influence on Gunung Binaia, Seram, Indonesia (Edwards et al. 1990; Edwards, Proctor, and Riswan 1993; Payton 1993).

Although limestone can profoundly influence vegetation, no generalizations are possible. Anderson (1965) has commented that in no other vegetation types in Sarawak is there such variation. One might question, in view of this variation, if there is justification for recognizing forest over limestone as a formation (the same applies to forest over ultramafic rocks) equal in rank to mangroves and other forest types, as Whitmore (1984) has done. Some earlier generalizations are simply not true: Anderson (1965) claimed that dipterocarps were rare or absent on limestone and that limestones were very species-rich. Yet dipterocarps accounted for 47.4% of the basal area of the plot on forest over limestone on Gunung Api, which was by far the least species-rich (for trees $\geq$ 10 cm dbh) of four plots on contrasting soils investigated by Proctor et al. (1983). It was, however, the most floristically distinct of these four plots and had proportionally many fewer species in common with the other three plots than these three plots had among themselves (Vallack 1982).

It is well known that many plant species are restricted to limestone-derived soils, and it seems likely (from analogies with studies in the temperate zone, such as Marrs and Bannister 1978) that where species occur both on and off this substratum, special races will have evolved for limestone soil. Many tropical limestones are thus reservoirs of genetic diversity; they should not be exploited without due consideration of their importance for conservation. If concern about the fragility of tropical forest soils applies anywhere, then surely it is on the limestones of Sarawak's Gunung Api, where natural fires have destroyed both vegetation and soils in some areas (Collins et al. 1984). These areas show no signs of recovery after many years, and if fires were to become more widespread, irreparable large-scale losses of forest and soils would occur.

Major advances in the understanding of temperate limestones have been made because of the ease of experimental work on herbaceous plants (such as Rorison and Robinson 1984). There is no reason why such work should not be attempted on herbaceous plants on tropical limestones. The observations by Burtt (1978) and others on various species in the Gesneriaceae are relevant. For instance, *Cyrtandra incrustata* is a Bornean plant found on limestone cliff faces. It accumulates calcium and has developed special glands to excrete this element, which causes heavy encrustation of calcium carbonate on stems and the undersides of leaves. *Cyrtandra calciphila*, in the same genus, grows on the same limestone cliffs as *C. incrustata* but lacks chalk glands. Clearly there has evolved within the genus at least two contrasting ways of tolerating the exigencies of the limestone habitat. Burtt and Tan (1984) made similar observations for the genus *Paraboea*. *Paraboea acutifolia* also has calcium-excreting glands. *Paraboea caerulescens* and *P. verticillata,* from Peninsular Malaya, and *P. effusa* and *P. meiophylla*, from Sarawak, however, have leaf blisters caused by the fusion of necrotic calcium accumulator cells. Within two genera of the Gesneriaceae there are thus at least three responses to high levels of calcium. These examples presumably reflect the wealth of physiological adaptations to be discovered in other species of forests over limestone.

Proctor (1992b) has reviewed forests over ultramafic rocks in Southeast Asia. The best documented are in New Caledonia, where large numbers of endemic species and some vegetation of strange physiognomy are restricted to ultramafics (Jaffré 1980). The plants include some extraordinary nickel-accumulating species, such as *Sebertia acuminata* (Sapotaceae), which has a blue-green latex compound of up to 25.7% dry weight of nickel (Jaffré et al. 1976). The ultramafic vegetation of Gunung Silam in Sabah (Proctor et al. 1988, 1989) has few endemics (unexpectedly so, in view of the general species-richness of Borneo) and just one nickel-accumulating species (and that a slight one—*Shorea tenuiramulosa* [Dipterocarpaceae]). There is evidence from Gunung Silam, however, of a subtle nickel-excretion mechanism operating via the litterfall (Proctor et al. 1989). Some magnificent examples of vegetation on ultramafic rocks occur in the Philippines, such as the extraordinarily stunted vegetation on Mt. Bloomfield, Palawan (J. Proctor et al., unpublished data, 1992). Here the ultramafic rock had a dramatic effect on the species composition, with few species able to cross the geological boundary on to the adjacent soils on sedimentary rocks. The sedimentary rocks had a forest in which the Symplocaceae, Myrtaceae, Dipterocarpaceae, and Thymeleaceae were the leading families (in order of their proportional contribution to the basal area), whereas a few meters away on the ultramafic soils the Myrtaceae (different species from those on the nonultramafic), Lauraceae, Chrysobalanaceae, and Ebenaceae were the most important. Baker et al. (1992) have reported from Mt. Bloomfield a new species of *Phyllanthus* (Euphorbiaceae) that had green sap and was another extraordinary nickel accumulator. In spite of the evidence for nickel toxicity at this site, extreme stunting seems to be caused at least in part by drought in the very shallow soils. Deeper ultramafic soils have larger forest growth, though still smaller in stature and floristically distinct from that on nonultramafic soils (Proctor and Nagy 1992). Sadly, the forests near the geological boundary on Mt. Bloomfield have recently been severely damaged by human activity. This highlights the plight of these unusual forests—unless described scientifically, they are easily overlooked for conservation, with priceless plants lost to science. Native plants of ultramafic soils, because of their clear-cut specializations to such an unusual chemical environment, offer the possibility of investigations ranging from the ecological to the molecular level of the nature of edaphic adaptations in tropical plant species. These investigations would almost certainly provide results of general importance.

REFERENCES

Alexander, I. J. 1989. Mycorrhizas in tropical forests. In *Mineral Nutrients in Tropical Forest and Savanna Ecosystems*. J. Proctor, editor. Blackwell, Oxford.

Anderson, J. A. R. 1965. The limestone habitat in Sarawak. In *Symposium on Ecological Research in Humid Tropical Vegetation*. J. A. R. Anderson, editor. UNESCO, Kuching, Sarawak.

Anderson, J. M., and T. Spencer. 1991. Carbon, nutrient and water balances of tropical rain forest ecosystems subject to disturbance: Management implications and research proposals. UNESCO, Paris, MAB Digest 7.

Ash, J. 1988. The location and stability of rainforest boundaries in north-eastern Queensland, Australia. *J. Biogeog.* 15: 619–630.

Ashton, P. S. 1976. Mixed dipterocarp forest and its variation with habitat in the Malayan lowlands: A re-evaluation at Pasoh. *Malay. For.* 39: 56–72.

Ashton, P. S., and P. Hall. 1992. Comparisons of structure amongst mixed dipterocarp forests of north-western Borneo. *J. Ecol.* 80: 459–481.

Austin, M. P., P. S. Ashton, and P. Greig-Smith. 1972. The application of quantitative methods in vegetative survey, III: A re-examination of rain forest data from Brunei. *J. Ecol.* 60: 309–324.

Baillie, I. C. 1989. Soil characteristics and classification in relation to the mineral nutrition of tropical wooded ecosystems. In *Mineral Nutrients in Tropical Forest and Savanna Ecosystems*. J. Proctor, editor. Blackwell, Oxford.

Baillie, I. C., and P. S. Ashton. 1983. Some soil aspects of the nutrient cycle in mixed dipterocarp forests in Sarawak. In *Tropical Rain Forest: Ecology and Management*. S. L. Sutton, T. C. Whitmore, and A. C. Chadwick, editors. Blackwell, Oxford.

Baillie, I. C., P. S. Ashton, M. M. Court, J. A. R. Anderson, E. A. Fitzpatrick, and J. Tinsley. 1987. Site characteristics and the distribution of tree species in mixed dipterocarp forest on Tertiary sediments in Central Sarawak, Malaysia. *J. Trop. Ecol.* 3: 201–220.

Baker, A. J. M., J. Proctor, M. M. J. van Balgooy, and R. D. Reeves. 1992. Hyper-accumulation of nickel by the flora of the ultramafics of Palawan, Republic of the Philippines. In *The Vegetation of Ultramafic (Serpentine) Soils: Proceedings of the First International Conference on Serpentine Ecology*. A. J. M. Baker, J. Proctor, and R. D. Reeves, editors. Intercept, Andover, England.

Baylis, G. T. S. 1975. The magnolioid mycorrhiza and mycotrophy in root systems derived from it. In *Endomycorrhizas*. F. E. Sanders, B. Mosse, and P. B. Tinker, editors. Academic Press, London.

Boxman, O., N. R. de Grauf, J. Hendrison, W. B. J. Jonkers, R. L. H. Poels, P. Schmidt, and R. L. S. Tijon. 1985. Towards sustained timber production from tropical rain forests in Suriname. *Neth. J. Agri. Sci.* 33: 125–132.

Bruijnzeel, L. A. 1991. *Hydrology of Moist Tropical Forests and Effects of Conversion: A State of Knowledge Review*. Free University, Amsterdam.

———. 1992. Managing tropical forest watersheds for production: where contradictory theory and practice co-exist. In *Wise Management of Tropical Forests*. F. R. Miller and K. L. Adam, editors. Oxford Forestry Institute, Oxford.

Bruijnzeel, L. A., M. L. Waterloo, J. Proctor, A. J. Kuiters, and B. Kotterink. 1993. Hydrological observations in montane rain forests on Gunung Silam, Sabah, Malaysia, with special reference to the "Massenerhebung" effect. *J. Ecol.* 81: 145–167.

Burnham, C. P. 1989. Pedological processes and nutrient supply from parent materials in tropical soils. In *Mineral Nutrients in Tropical Forest and Savanna Ecosystems*. J. Proctor, editor. Blackwell Scientific Publishers, Oxford.

Burtt, B. L. 1978. Studies in the Gesneriaceae of the Old World XLIV: New and little-known species of *Cyrtandra*, chiefly from Sarawak. *Notes Roy. Bot. Gard. Edinb.* 36: 157–179.

Burtt, B. L., and K. Tan. 1984. Studies in the Gesneriaceae of the Old World XLVIII: Calcium accumulation and excretion in *Paraboea*. *Notes Roy. Bot. Gard. Edinb.* 41: 453–456.

Chin, S. C. 1977. The limestone hill flora of Malaya I. *Gard. Bull. Singapore* 30: 165–220.

Collins, N. M., J. D. Holloway, and J. Proctor. 1984. Notes on the ascent and natural history of Gunung Api, a limestone mountain in Sarawak. *Sarawak Mus. J.* 33: 219–234.

Crowther, J. 1982. Ecological observations in a tropical karst terrain, West Malaysia, I: Variations in topography, soils and vegetation. *J. Biogeog.* 65–78.

Denslow, J. S., P. M. Vitousek, and J. C. Schultz. 1987. Bioassays of nutrient limitation in a tropical rain forest soil. *Oecologia* (Berlin) 74: 370–376.

Edwards, I. D., R. W. Payton, J. Proctor, and S. Riswan 1990. Altitudinal zonation of the rain forests in Manusela National Park, Seram, Maluku, Indonesia. In *The Plant Diversity of Malaysia*. P. Baas and H. P. Nooteboom, editors. Kluwer, Dordrecht, The Netherlands.

Edwards, I. D., J. Proctor, and S. Riswan 1993. Rain forest types in the Manusela National Park. In *Natural History of Seram, Maluku, Indonesia*. I. D. Edwards, A. A. MacDonald, and J. Proctor, editors. Intercept, Andover, England.

Evans, J. 1992. *Plantation Forestry in the Tropics*. Second edition. Clarendon Press, Oxford.

Fitter, A. H. 1986. Effect of benomyl on leaf phosphorus concentration in alpine grasslands: a test of mycorrhizal benefit. *New Phytol.* 103: 53–56.

———. 1991. Costs and benefits of mycorrhizas: Implications for functioning under natural conditions. *Experientia* 47: 350–55.

Furley, P. A., J. A. Ratter, and D. R. Gifford. 1988. Observations on the vegetation of eastern Mato Grosso, Brazil, III: The woody vegetation of the Morro de Fumaca, Torixoreu. *Proc. Roy. Soc. Lon.* B 235: 259–280.

Gerrish, G., D. Mueller-Dombois, and K. W. Bridges. 1988. Nutrient limitation and *Metrosideros* forest dieback in Hawaii. *Ecology* 69: 723–727.

Gillman, G. P., D. F. Sinclair, R. Knowitan, and M. G. Keys. 1985. The effect on some soil chemical properties of the selective logging of a north Queensland rain forest. *For. Ecol. & Mgmt.* 12: 195–214.

Green, J. J. 1993. Fine root dynamics in a Bornean rain forest. Ph. D. diss., University of Stirling, Stirling, Scotland.

Grubb, P. J. 1989. The role of mineral nutrients in the tropics: A plant ecologist's view. In *Mineral Nutrients in Tropical Forest and Savanna Ecosystems*. J. Proctor, editor. Blackwell Scientific, Oxford.

Grubb, P. J., and P. J. Edwards. 1982. Studies in mineral cycling in a montane rain forest in New Guinea, III: The distribution of mineral elements in the above-ground material. *J. Ecol.* 70: 623–648.

Hayman, D. S. 1987. VA mycorrhizas in field crop systems. In *Ecophysiology of VA Mycorrhizal Plants*. G. R. Safir, editor. CRC Press, Florida.

Heaney, A., and J. Proctor. 1989. Chemical elements in litter on Volcan Barva, Costa Rica. In *Mineral Nutrients in Tropical Forest and Savanna Ecosystems*. J. Proctor, editor. Blackwell Scientific, Oxford.

Ho, C. C., D. M. Newbery, and M. E. D. Poore. 1987. Forest composition and inferred dynamics in Jengka Forest Reserve, Malaysia. *J. Trop. Ecol.* 3: 25–56.

Jaffré, T. 1980. *Etude écologique du peuplement végétal des sols dérivés de roches ultrabasiques en Nouvelle Calédonie*. ORSTOM, Paris.

Jaffré, T. R. R. Brooks, J. Lee, and R. D. Reeves. 1976. *Sebertia acuminata:* A hyperaccumulator of nickel from New Caledonia. *Science* 193: 579–580.

Janos, D. P. 1987. VA mycorrhizas in humid tropical ecosystems. In *Ecophysiology of VA Mycorrhizal Plants*. G. R. Safir, editor. CRC Press, Florida.

Jonkers, W. B. J. 1989. *Vegetation Structure, Logging Damage and Silviculture in a Tropical Rain Forest in Suriname*. Agricultural University, Wageningen, The Netherlands.

Jordan, C. F. 1985. *Nutrient Cycling in Tropical Forest Ecosystems*. John Wiley, New York.

Kruckeberg, A. R. 1992. Plant life of western North American ultramafics. In *The Ecology of Areas with Serpentized Rocks: A World View*. B. A. Roberts and J. Proctor, editors. Kluwer, Dordrecht, The Netherlands.

Marrs, R. H., and P. Bannister. 1978. The adaption of *Calluna vulgaris* (L.) Hull to contrasting soil types. *New Phytol.* 81: 753–761.

Marrs, R. H., J. Proctor, A. Heaney, and M. D. Mountford. 1988. Changes in soil nitrogen mineralization and nutrification along an altitudinal transect in tropical rain forest in Costa Rica. *J. Ecol.* 76: 466–482.

Marrs, R. H., J. Thompson, D. A. Scott, and J. Proctor. 1991. Nitrogen mineralization and nutrification in *terra firme* forest and savanna soils on Ilha de Maracá, Roraima, Brazil. *J. Trop. Ecol.* 7: 123–137.

McGonigle, T. P. 1988. A numerical analysis of published field trials with vesicular-arbuscular mycorrhizal fungi. *Funct. Ecol.* 2: 473–478.

Nemec, S. 1987. VA mycorrhizas in horticultural systems. In *Ecophysiology of VA Mycorrhizal Plants*. G. R. Safir, editor. CRC Press, Florida.

Newbery, D. M., and J. Proctor. 1984. Ecological studies in four contrasting lowland rain forests in Gunung Mulu National Park, Sarawak. *J. Ecol.* 72: 475–493.

Payton, R. W. 1993. Soils of the Manusela National Park. In *Natural History of Seram, Maluku, Indonesia*. I. D. Edwards, A. A. MacDonald, and J. Proctor, editors. Intercept, Andover, England.

Poels, R. 1987. *Soils, Water and Nutrients in a Forest Ecosystem in Suriname*. Agricultural University, Wageningen, The Netherlands.

Primack, R. B., E. O. K. Chai, S. S. Tan, and H. S. Lee. 1987. The silviculture of dipterocarp trees in Sarawak, Malaysia, I: Introduction to the series and performance in primary forest. *Malay. For.* 50: 29–42.

Proctor, J. 1971. The plant ecology of serpentine, III: The influence of a high Mg/Ca ratio and high nickel and chromium levels in some British and Swedish serpentine soils. *J. Ecol.* 59: 827–842.

————. 1983. Mineral nutrients in tropical forests. *Prog. Phys. Geog.* 7: 422–431.

————. 1987. Nutrient cycling in primary and old secondary rain forests. *Appl. Geog.* 7: 135–152.

————. 1992a. Soils and mineral nutrients: What do we know and what do we need to know for wise rain forest management? In *Wise Management of Tropical Forests: Proceedings of the Oxford Conference on Tropical Forests, 1992*. F. R. Miller and K. L. Adam, editors. Oxford Forestry Institute, Oxford.

————. 1992b. The vegetation over ultramafic rocks in the tropical far east. In *The Ecology of Areas with Serpentized Rocks: A World View*. B. A. Roberts and J. Proctor, editors. Kluwer, Dordrecht, The Netherlands.

Proctor, J., J. M. Anderson, P. Chai, and H. W. Wallak. 1983. Ecological studies in four contrasting lowland forests in Gunung Mulu National Park, Sarawak, I: Forest environment, structure and floristics. *J. Ecol.* 71: 237–260.

Proctor, J., Y. F. Lee, A. M. Langley, W. R. C. Munro, and T. Nelson. 1988. Ecological studies on Gunung Silam, a small ultrabasic mountain in Sabah, Malaysia, I: Environment, forest structure and floristics. *J. Ecol.* 76: 320–340.

Proctor, J., and L. Nagy. 1992. The ecology of serpentine soils: An overview. In *The Ecology of Ultramafic (Serpentine) Soils*. Intercept, London.

Proctor, J. C., Phillips, G. K. Duff, A. Heaney, and F. M. Robertson. 1989. Ecological studies on Gunung Silam, a small ultrabasic mountain in Sabah, Malaysia, II: Environment, forest structure and floristics. *J. Ecol.* 77: 317–331.

Rampanana, L., J. L. Rakotomanana, D. Louppe, and F. Brunck. 1988. Dying of *Pinus kesiya* crowns in Madagascar. *Bois & For. Trop.* 214: 23–47.

Rance, S. J., D. M. Cameron, and E. R. Williams. 1982. Correction of crown disorders of *Pinus carribaea* var. *hondurensis* by application of zinc. *Plant & Soil* 65: 293–296.

Ratter, J. A., G. P. Askew, R. F. Montgomery, and D. R. Gifford. 1978. Observations on forests of some mesotrophic soils in central Brazil. *Rev. Bras. Bot.* 1: 47–58.

Ratter, J. A., P. W. Richards, G. Argent, and D. R. Gifford. 1973. Observations on the vegetation of northeastern Mato Grosso, I: The woody vegetation types of the Xavantina-Cachimbo Expedition area. *Phil. Trans. Roy. Soc. Lon.* B 206: 449–492.

Roberts, B. A., and J. Proctor, editors. 1992. *The Ecology of Areas with Serpentized Rocks: A World View*. Kluwer, Dordrecht, The Netherlands.

Rorison, L. H., and D. Robinson. 1984. Calcium as an environmental barrier. *Plant, Cell & Environ.* 7: 381–391.

Santoso, E. 1988. Pengaruh mikoriza terhadap diameter batang da bobot kering anakan Dipterocarpaceae. *Bull. Pen. Hutan* 504: 11–21.

Scott, D. A., J. Proctor, and J. Thompson. 1992. Ecological studies on a lowland evergreen rain forest on Maracá Island, Roraima, Brazil, II: Litter and nutrient cycling. *J. Ecol.* 80: 705–717.

Tanner, E. V. J., V. Kapos, and W. Franco. 1992. Nitrogen and phosphorus fertilization effects on Venezuelan montane forest trunk growth and litterfall. *Ecology* 73: 78–86.

Tanner, E. V. J., V. Kapos, S. Freskos, J. R. Healey, and A. N. Theobald. 1990. Nitrogen and phosphorus fertilization of Jamaican montane forest trees. *J. Trop. Ecol.* 6: 231–238.

Thompson, J., J. Proctor, W. Milliken, J. A. Ratter, and D. A. Scott. 1992a. The forest-savanna boundary on Maracá Island, Roraima, Brazil: An investigation of two contrasting transects. In *Nature and Dynamics of Forest-Savanna Boundaries*. J. Proctor and J. A. Ratter, editors. Chapman and Hall, London.

Thompson, J., J. Proctor, and D. A. Scott. 1994. A semi-evergreen forest on Maracá Island I: Physical environment, forest structure and floristics. In *The Rainforest Edge*, J. Hemming, editor. Manchester University Press, Manchester.

Thompson, J., J. Proctor, V. Viana, W. Milliken, J. A. Ratter, and D. A. Scott. 1992b. Ecological studies on a lowland evergreen rain forest on Maracá Island, Roraima, Brazil, I: Physical environment, forest structure and leaf chemistry. *J. Ecol.* 80: 689–703.

Turner, I. M. 1991. Effects of shade and fertilizer addition on the seedlings of two tropical woody pioneer species. *Trop. Ecol.* 32: 24–29.

Turner, I. M., N. D. Brown, and A. C. Newton. 1993. The effect of fertilizer application on dipterocarp seedling growth and mycorrhizal infection. *For. Ecol. Mgmt.* 57: 329–337.

Vallack, H. W. 1981. Ecological studies in a tropical rain forest site on limestone in Gunung Mulu National Park, Sarawak. M.Sc. thesis, University of Stirling, Stirling, Scotland.

Vitousek, P. M., and R. L. Sanford. 1987. Nutrient cycling in moist tropical forest. *Ann. Rev. Ecol. Sys.* 17: 137–167.

Vitousek, P. M., L. R. Walker, L. D. Whitaker, D. Mueller-Dombois, and P. A. Matson. 1987. Biological invasion by *Myrica fava* alters ecosystem development in Hawaii. *Science* 238: 802–804.

Whitmore, T. C. 1984. *Tropical Rain Forests of the Far East.* Clarendon Press, Oxford.

Wong, Y. K., and T. C. Whitmore. 1970. On the influence of soil properties on species distribution in a Malayan lowland dipterocarp forest. *Malay. For.* 33: 42–54.

EFFECTS OF SELECTIVE LOGGING ON SOIL CHARACTERISTICS AND GROWTH OF PLANTED DIPTEROCARP SEEDLINGS IN SABAH

Ruth Nussbaum, Jo Anderson, and Tom Spencer

The devastation caused by clearing huge areas of rainforest has been well documented and widely publicized (for example, Sanchez 1976). One factor contributing to poor recovery of the residual forest is the reduction in soil fertility associated with excessive damage to the soil (Lal 1987). How much commercial logging contributes to this degradation is controversial; much information about the effects of changes in soil properties on forest recovery comes from studies of other types of disturbance, such as slash-and-burn agriculture or large-scale clearing and burning for agriculture and pasture. Nevertheless, there is increasing pressure to reduce logging damage in rainforests to a level that does not exceed the forests' capacity to recover. To determine the effects on forest recovery of factors directly related to logging activities, such as soil compaction, loss of topsoil and organic matter, and the creation of openings in the canopy and debris piles, we have been studying sites in the Ulu Segama Forest Reserve (USFR) in eastern Sabah, Malaysia.

SELECTIVE LOGGING

Timber harvesting occurs on a large scale throughout much of Southeast Asia. Logging practices in Sabah are fairly typical of forest exploitation in the region. Timber is harvested using a selective system in which all commercial timber trees with a diameter at breast height (dbh) greater than 60 cm are harvested in a single operation (Poore 1989). In comparison with other mixed dipterocarp forests in Southeast Asia, Sabah is heavily stocked with timber trees (for stocking figures see Manokaran and Kochummen 1987; Newbery et al. 1992). Consequently, extraction rates are relatively high (about 8–14 trees/ha^{-1}).

Logging operations create a mosaic of disturbed areas. The roads used for general access, the log-landing areas where logs are collected and loaded onto trucks, and the skid trails made by bulldozers are all completely cleared of vegetation, litter, and often topsoil. Where trees fall, the closed canopy is replaced by gaps and large piles of debris. Elsewhere, the disturbance may be a relatively small increase in debris or light levels; other forest patches remain undisturbed.

To study the effects of disturbance following selective logging, we classified logged areas, based on visual observation of damage to vegetation and soil, into the following five classes:

1. Log-landing areas: all vegetation, litter, and topsoil are removed.
2. Skid trails: all vegetation, litter, and usually topsoil are removed.
3. Debris piles: the canopy is disturbed, the understory is often broken or damaged, and debris covers the soil surface, but the litter layer and topsoil are mostly undisturbed.
4. Disturbed forest: the canopy is somewhat disturbed, and there is some damage to trees; however, there is usually little disturbance to understory, little or no debris, and no disturbance to litter and topsoil.
5. Undisturbed forest: there is no disturbance to canopy, understory, litter, or topsoil related to logging operations.

A survey of 300 ha of logged forest in the USFR indicated that log landings and skid trails account for approximately 30% of logged land areas (log landings 5% and skids 25%), debris piles account for 30%, disturbed forest 20%, and residual undisturbed patches make up 20% of the land area (R. Nussbaum, unpublished data, 1992).

SOIL COMPACTION

Several studies have shown that forest clearance often results in soil compaction. Soil bulk density increases (Lal 1987), while pore volume and infiltration rates decrease (Chauvel, Grimaldi, and Tessier 1991). Both soil type and moisture content during disturbance influence the degree of compaction (Dias and Nortcliff 1985a; Malmer and Grip 1990). The method of clearing is also an important variable, as studies of large-scale clearing in Brazil (Chauvel, Grimaldi, and Tessier 1991; Dias and Nortcliff 1985b), Sabah (Malmer and Grip 1990), Peru (Seubert, Sanchez, and Valverde 1977), and Suriname (Van der Weert 1974) indicate that heavy machinery is the main cause of soil compaction. Our findings concur with these results.

We worked in 25 ha of recently logged forest in which skid trails and log landings had been substantially disturbed by heavy machinery (fig. 7.1). Soil bulk density in these areas was significantly greater than in areas that had not been traversed by heavy machinery (debris piles and disturbed forest) (table 7.1). Log-landing soils were significantly more compacted than the skid trails. This

FIGURE 7.1 Logs being dragged out of the forest. Note the severe erosion along the sides of the trail. Photograph by R. Primack.

Table 7.1 Physical and chemical properties of soils at two depths in areas differently disturbed by selective logging in eastern Sabah

Disturbance class	Bulk density (g cm⁻¹)	pH (H₂O)		Organic carbon (%)		Total P (mg kg⁻¹)		Total N (%)	
	0–5 cm	0–5 cm	5–15 cm	0–5 cm	5–15 cm	0–5 cm	5–15 cm	0–5 cm	5–15 cm
Log landings	1.66	5.07	4.91	1.31	0.53	106.1	103.1	0.08	0.08
Skid trails	1.44	4.99	5.04	1.52	0.98	163.6	150.1	0.15	0.13
Debris piles	1.01	5.04	4.90	3.13	1.49	245.8	184.3	0.26	0.16
Disturbed forests	1.06	4.83	4.85	2.71	1.40	208.1	164.6	0.24	0.14
Undisturbed forests	1.08	4.54	4.75	3.05	1.19	225.4	174.4	0.25	0.16
Least standard deviation ($p < 0.05$)	0.15	0.325		0.495		23.18		0.036	

Note: All values are means of approximately twenty samples.

observation may result from the loss of topsoil, which left the more compacted subsoil exposed, or from the intensive use of heavy machinery, although Dias and Nortcliff (1985a) reported that most of the damage to soil structure occurs during the first one or two passes of a bulldozer.

EROSION

Erosion rates in undisturbed rainforest are generally very low (Lal 1986) but increase, often dramatically, after logging. Measurements of the suspended sediment load of streams draining from logged areas have shown that soil losses are usually large immediately after logging and decrease gradually over time (Gilmour 1977). In the USFR, suspended sediment yields from a logged catchment were 18 times higher than those from an adjacent unlogged catchment during the first five months following logging, dropping to 3.6 times higher after one year (Douglas et al. 1992). The sources of eroded soil, however, are not distributed evenly over the entire logged area. Erosion rates in undisturbed forest fragments are unlikely to increase significantly (Lal 1986; Sinun 1991). Most eroded soil originates from such areas as log landings and skid trails, where loss of the protective litter layer and herbaceous cover allows the direct impact of raindrops to detach soil particles. These particles are transported by overland flow, the volume of which is substantially increased by reduced rainfall infiltration into the soil. Stated simply, instead of quickly soaking into the leaf-covered soil, the rainwater runs over the surface and carries away the exposed soil particles. Infiltration rates on log landings in western Sabah were found to be as low as 0.58 mm/hr^{-1}, compared to 154 mm/hr^{-1} in undisturbed forest (Malmer and Grip 1990), whereas in the USFR no measurable infiltration had occurred on log landings after two hours.

To study the process of erosion on bare soil in more detail, we established a network of about 300 erosion pins on the log landing and skid trails of a 25 ha logged area. An average of 21.6 mm of soil was lost per pin from the log landing over the first year, and 11.9 mm was lost per pin from skid trails. Soil loss is not a simple process, however, but the result of a series of small erosion and deposition events. Approximately 80% of sites showed net erosion over the first year, but 20% showed net deposition. Deposition occurred mainly where sediment was trapped behind organic debris, such as bark, or in natural hollows. This finding has important implications for management, as it may be possible to increase the number of deposition sites to reduce overall erosion. An extensive gully network, which appeared to channel much of the runoff and sediment directly into streams, formed on the log landing and skids in less than two months after logging. Therefore, it may be possible to reduce loss of soil from a catchment by using cross drains or water bars to divert overland flow from skids and log landings into adjacent areas with undamaged soil.

SOIL NUTRIENTS AND ORGANIC MATTER

Nutrients in rainforest soils are part of a complex system of nutrient cycling that encompasses the growth and decomposition of plant and animal biomass, hydrology, activity of soil macro- and microflora, losses through leaching and denitrification, and gains through weathering of rocks and wet and dry deposition. The balanced nutrient cycle of a mature forest can be drastically altered by disturbance. Some forests grow on extremely nutrient-poor soils with very low external nutrient inputs; if nutrients are lost, these soils become infertile. Other forests grow on more fertile soil; lost nutrients can be replenished gradually from weathering of bedrock, inputs in precipitation, and nitrogen fixation. These soils are better able to withstand disturbance, although fertility may be irreversibly lost if the impact of disturbance is too severe (Burnham 1989; Jordan and Herrera 1981; Vitousek and Sanford 1986). In some areas, soil nutrients were so depleted after large-scale forest clearing that the land was rapidly abandoned as useless (Sanchez 1976). If selectively logged forests are to retain their productive capacity, their soils must be managed to maintain sufficient nutrient capital for forest regeneration.

The topsoil contains a high proportion of organic matter and nutrients (Lal 1987). For this reason, rainforest soils may lose large amounts of nutrients and organic matter with the removal of only a few centimeters of soil. During logging operations, the topsoil on skid trails and log landings is frequently pushed to one side by bulldozers (Gillman et al. 1985; Nussbaum, personal observation, 1991). We observed that the concentrations of organic matter, nitrogen, and phosphorus on skid trails and log landings were extremely low three months after logging, and soil acidity was also somewhat reduced (see table 7.1). The top 5 cm of soil had slightly higher concentrations than the underlying soil, probably from organic debris mixed into the soil by the bulldozers during logging. These top few centimeters, however, were partially eroded during the following year.

Although nutrients may be released by decomposing organic debris (Lal 1987), we found that concentrations of nitrogen, phosphorus, and organic matter in soils under debris piles were similar to those of undisturbed areas. Nutrients were possibly released but not detected because of the speed with which they are taken up by regrowing vegetation or leached by rainfall (Vitousek and Denslow 1986).

The soils in logged forest fall into two broad categories: degraded soils, which have been traversed by heavy machinery, and nondegraded soils, which have not. This dichotomy is likely to have implications for forest recovery, especially with respect to the growth of dipterocarp seedlings, which are of particular importance in Southeast Asian forests because dipterocarps are the principal timber trees.

Soil properties in disturbed areas not traversed by heavy machinery are very similar to those in undisturbed forest patches. The major changes to these areas are vegetation damage and an associated decrease in canopy cover, resulting in an increase in light intensity. Many authors have reported enhanced growth of dipterocarp seedlings in higher light intensities (Ang 1991; Sasaki and Mori 1981; Whitmore 1984). If canopy openings are too large, however, the high light intensities may promote the rapid growth of light-demanding vines and climbing bamboos, which smother dipterocarp seedlings and other vegetation (Chai 1975; Mead 1937; Priasukmana 1989; Tang and Wadley 1976). We observed this pattern on our study site, where seedlings of three dipterocarp species (*Dryobalanops lanceolata* [n = 45], *Parashorea malaanonan* [n = 120] and *Shorea parvifolia* [n = 52]) were planted four months after logging (fig. 7.2). Canopy

FIGURE 7.2   Naturally occurring kapur (*Dryobalanops*) seedling being measured for height on the forest floor. Note the plastic label at the base of the seedling, where the old wings of the fruit can still be seen. Photograph by R. Primack.

Table 7.2   Mean height increments of three species of dipterocarp seedlings planted in different disturbance classes in recently logged forests in eastern Sabah

	Dryobalanops lanceolata	*Parashorea malaanonan*	*Shorea parvifolia*
Log landings	—	−10.5	11.8
Skid trails	46.6	14.3	31.1
Debris piles	70.6	63.1	73.2
Disturbed forests	77.6	36.4	—
Undisturbed forests	52.6	35.1	55.8
Least standard deviation ($p < 0.05$)	29.3	24.7	40.7

Note: Values indicate the height increment in cm of seedlings after 12 months.

opening correlated positively with height increment of seedlings planted in undisturbed soil (debris piles, disturbed forest, and undisturbed fragments) (table 7.2). In areas where less than 50% of the original canopy remained, up to 73% of seedlings were overgrown by vines, compared to only 11% in sites with undisturbed canopy.

Soil on skid trails and log landings generally is compacted and nutrient-poor with accelerated rates of erosion; few or no living plants remain. Decades after logging operations are completed, such areas show very poor recovery of soil properties and vegetation (Glauner, personal communication, 1991; Malmer and Grip 1990) and may take up to a thousand years to recover their original biomass (Uhl et al. 1982). A variety of factors may contribute to poor recovery: initial absence of seeds (Pinard, Howlett, and Davidson 1994) or seedlings, inhibition of root growth from soil compaction (Van der Weert 1974), reduced infiltration and low water-holding capacity leading to water stress and increased mortality (Awang and Sawal 1986), high soil temperatures with damaging effects on mycorrhizae (Lee 1988; Smits 1983), high leaf temperatures (Ashton and De Zoysa 1989; Kamaluddin and Grace 1992; Sasaki and Mori 1981), and increased herbivory by mammals (Becker 1985; Wyatt-Smith 1963) and insects (Becker 1983; Daljeet-Singh 1975). In our study, the average height increment of seedlings planted on skid trails was significantly lower than that of seedlings planted in sites with undisturbed soil but similar canopy opening. Seedlings planted on the log landing had a lower height increment than seedlings in any other site.

Poor recovery on log landings and skid trails results in the loss of substantial areas of potentially productive forest. In addition, these areas provide a habitat for vines and climbing bamboos, which spread into the surroun-

Table 7.3 Relative height increment and total dry weight of six-month-old seedlings of one dipterocarp species (*Shorea leprosula*) and one pioneer tree (*Macaranga gigantea*) planted in six soil treatments on log landings (*n* = 12)

| | Relative height increment | | Dry weight (g) | |
	Shorea	*Macaranga*	*Shorea*	*Macaranga*
Control	0.35	0.70	3.67	4.18
Fertilizer	0.94	2.63	27.36	151.76
Mulch	0.41	1.69	8.67	18.19
Plowing	0.51	2.17	9.62	28.64
Plowing and fertilizer	1.22	4.09	47.24	249.55
Plowing and mulch	0.47	0.88	8.10	15.61
Topsoil	1.12	3.65	42.24	227.93
Standard error	0.15	5.90	0.54	27.25

ding forest, further slowing recovery. To investigate the importance of soil amelioration in the rehabilitation of such areas, we tested the effects of several soil treatments on the growth of dipterocarp and local pioneer tree seedlings (table 7.3): (1) plowing to reduce compaction, (2) fertilizing to replace nutrients, (3) mulching to reduce high soil temperatures and water loss, (4) a combination of plowing and fertilizing, (5) a combination of plowing and mulching, and (6) replacing topsoil (Nussbaum, Anderson, and Spencer, in press). The treatment that most markedly promoted rapid early growth was the application of fertilizer (40 g NPK/plant). Reduced compaction from plowing also resulted in enhanced growth, particularly in combination with fertilizer application. The height and crown diameter increments of pioneer tree seedlings were up to seven times greater than those of the dipterocarps. Rapid growth of pioneer tree seedlings may allow them to compete with vines and protect the soil from erosion.

CONCLUSION

Soil characteristics do not change significantly following timber harvesting in areas in which heavy machinery was not used. Most of the damage to the soil, including compaction, accelerated erosion, and loss of nutrients and organic matter, occurs on skid trails and log landings. Compaction, the result of repeated use of heavy machinery together with topsoil removal and exposure of more compacted subsoil, is less severe on skid trails than on log landings, but it is sufficient to inhibit plant growth in both cases. Reduced rainfall infiltration on compacted soils, together with poor recovery of litter and herbaceous cover, result in accelerated rates of erosion. Erosion depletes soil nutrients and organic matter, although the most serious loss occurs when topsoil is removed during logging operations. All of these factors contribute to a decline in soil fertility, which inhibits the recovery of vegetation.

The most effective way to control soil damage during logging is to minimize the area covered by log landings and skid trails. Improved management of logging operations can reduce the extent of such areas by more than half (Hendrison 1990; Nicholson 1979), and these management techniques are now being introduced in the Ulu Segama Forest Reserve. Throughout the tropics, however, many millions of hectares of rainforest have already been degraded by poor management. Restoration of degraded secondary forests will require some form of rehabilitation such as enrichment planting (Moura-Costa, this volume). Areas where the soil has been severely damaged may require efforts to prevent further degradation and restore soil fertility. It may be possible to reduce soil loss on degraded sites by diverting flows from compacted soil, thus increasing the frequency of deposition events. The establishment of protective layers of vegetation and litter is essential. Growth of vegetation can be encouraged by restoring soil fertility, using plowing to reduce compaction, and adding fertilizer to replace nutrients. Fast-growing indigenous pioneer trees may be particularly appropriate for revegetating such areas, as they outperform dipterocarp seedlings and respond well to soil amelioration treatments.

REFERENCES

Ang, L. H. 1991. Effects of open and under planting on early survival and growth of *Endospermum malaccense* (sesendok), *Alstonia angustiloba* (pulai) and *Shorea parvifolia* (meranti sarang punai). *J. Trop. For. Sci.* 3: 380–384.

Ashton, P. M. S., and N. D. De Zoysa. 1989. Performance of *Shorea trapezifolia* (Thwaites) Ashton seedlings growing in different light regimes. *J. Trop. For. Sci.* 1, 4: 356–364.

Awang, K., and P. Sawal. 1986. Initial performance of three indigenous species used in an enrichment planting of hill dipterocarp forest. In *Proceedings of the Workshop on Impact of Man's Activities on Upland Forest Ecosystems*. H. Yusuf, K. Awang, N. M. Majid, and S. Mohamed, editors. Universiti Pertanian Malaysia, Selangor.

Becker, P. 1983. Effects of insect herbivory and artificial defoliation on survival of *Shorea* seedlings. In *Tropical Rainforest Ecology and Management*. S. L. Sutton, T. C. Whitmore, and A. C. Chadwick, editors. Blackwell Scientific, Oxford.

———. 1985. Catastrophic mortality of *Shorea leprosula* juveniles in a small gap. *Malay. For.* 48: 263–265.

Burnham, C. P. 1989. Pedological processes and nutrient supply from parent material in tropical soils. In *Tropical Forest and Savanna Ecosystems*. J. Proctor, editor. Special Publication No. 9 of the British Ecological Society. Blackwell Scientific, Oxford.

Chai, D. N. P. 1975. Enrichment planting in Sabah. *Malay. For.* 38: 271–277.

Chauvel, A., M. Grimaldi, and D. Tessier. 1991. Changes in soil pore-space distribution following deforestation and revegetation: An example from the Central Amazon Basin, Brazil. *For. Ecol. & Mgmt.* 38: 259–271.

Daljeet-Singh, K. 1975. A preliminary survey of insect attack on seedlings and saplings in Bukit Belata Forest Reserve. *Malay. For.* 38: 14–16.

Dias, A. C. C. P., and S. Nortcliff. 1985a. Effects of tractor passes on the physical properties of an oxisol in the Brasilian Amazon. *Trop. Agr.* (Trinidad) 62: 137–141.

———. 1985b. Effects of two land clearing methods on the physical properties of an oxisol in the Brasilian Amazon. *Trop. Agr.* (Trinidad) 62: 207–212.

Douglas, I., T. Spencer, T. Greer, K. Bidin, W. Sinun, and W. M. Wong. 1992. The impact of selective commercial logging on stream hydrology, chemistry and sediment loads in the Ulu Segama rain forest, Sabah, Malaysia. *Phil. Trans. Roy. Soc. Lon.* B 335: 397–406.

Gillman, G. P., D. F. Sinclair, R. Knowlton, and M. G. Keys. 1985. The effect on some soil chemical properties of the selective logging of a north Queensland rainforest. *For. Ecol. & Mgmt.* 12: 195–214.

Gilmour, D. A. 1977. Effects of rainforest logging and clearing on water yield and quality in a high rainfall zone of northeast Queensland. In *Hydrological Symposium of the Institute of Engineers.* National Conference Publication No. 77/5, The Institution of Engineers, Canberra.

Hendrison, J. 1990. *Damage-Controlled Logging in Managed Rain Forest in Suriname.* Agricultural University, Wageningen, The Netherlands.

Jordan, C. F., and R. Herrera. 1981. Tropical rainforests: Are nutrients really critical? *Am. Nat.* 117: 167–180.

Kamaluddin, M., and J. Grace. 1992. Photoinhibition and light acclimation in seedlings of *Bischofia javanica,* a tropical forest tree from Asia. *Ann. Bot.* 69: 47–52.

Lal, R. 1986. Deforestation and soil erosion. In *Land Clearing and Development in the Tropics.* R. Lal, P. A. Sanchez, and R. W. Cummings, editors. A. A. Balkema, Rotterdam, The Netherlands.

———. 1987. *Tropical Ecology and Physical Edaphology.* Wiley, London.

Lee, S. S. 1988. The ectomycorrhizas of *Shorea leprosula* (Dipterocarpaceae). In *Proceedings of the Asian Seminar on Trees and Mycorrhiza.* F. S. P. Ng, editor. FRIM, Kepong.

Malmer, A., and H. Grip. 1990. Soil disturbance and loss of infiltrability caused by mechanized and manual extractions of tropical rainforest in Sabah, Malaysia. *For. Ecol. & Mgmt.* 38: 1–12.

Manokaran, N., and K. M. Kochummen. 1987. Recruitment, growth and mortality of tree species in a lowland dipterocarp forest in Peninsular Malaysia. *J. Trop. Ecol.* 3: 315–330.

Mead, J. P. 1937. The silvicultural treatment of young meranti crops. *Malay. For.* 6: 53–56.

Moura-Costa, P. H. This volume. Nursery and vegetative propagation techniques for genetic improvement and large-scale enrichment planting of dipterocarps.

Newbery, D. M., E. J. F. Campbell, Y. F. Lee, C. E. Risdale, and M. J. Still. 1992. Primary lowland dipterocarp forest at Danum Valley, Sabah, Malaysia: Structure, relative abundance and family composition. *Phil. Trans. Roy. Soc. Lon.* B 335: 341–356.

Nicholson, D. I. 1979. The effects of logging and treatment on the mixed dipterocarp forests of South East Asia. FAO Misc. Ser. 79, FAO, Rome.

Nussbaum, R., J. Anderson, and T. Spencer. In press. Factors limiting the growth of indigenous tree seedlings planted on degraded forest soils in Sabah Malaysia. *For. Ecol. & Mgt.*

Pinard, M., B. Howlett, and D. Davidson. In press. Site conditions limit pioneer tree recruitment after logging of dipterocarp forests in Sabah, Malaysia. *Biotropica.*

Poore, D. 1989. *No Timber without Trees: Sustainability in the Tropical Forest.* Earthscan, London.

Priasukmana, S. 1989. Planting experiments of dipterocarps in East Kalimantan. In *Fourth Round-Table Conference on Dipterocarps.* I. Soerianegara, S. S. Tjitrosomo, R. C. Umaly, and I. Umboh, editors. Biotrop. Spec. Pub. 41.

Sanchez, P. A. 1976. *Properties and Management of Soils in the Tropics.* Wiley, New York.

Sasaki, S., and T. Mori. 1981. Growth responses of dipterocarp seedlings to light. *Malay. For.* 44: 319–345.

Seubert, C. E., P. A. Sanchez, and C. Valverde. 1977. Effects of land clearing methods on soil properties of an ultisol and crop performance in the Amazon jungle of Peru. *Trop. Agr.* 54: 307–321.

Sinun, W. 1991. Hillslope hydrology, hydrogeomorphology and hydrochemistry of an equatorial lowland rainforest, Sabah, Malaysia. M.A. thesis, University of Manchester, Manchester, England.

Smits, W. T. M. 1983. Dipterocarps and mycorrhiza: An ecological adaptation and a factor in forest regeneration. *Flora Malesiana* 36: 3926–3937.

Tang, H. T., and H. Wadley. 1976. Report on the survival and development survey of areas reforested by line-planting in Selangor. FRI Research Pamphlet No. 67, Forest Research Institute, Kepong, Kuala Lumpur.

Uhl, C., C. Jordan, K. Clark, H. Clark, and R. Herrera. 1982. Ecosystem recovery in Amazon caatinga forest after cutting, cutting and burning and bulldozer clearing treatments. *Oikos* 38: 313–320.

Van der Weert, A. 1974. The influence of mechanical forest clearing on soil conditions and resulting effects on root growth. *Trop. Agr.* (Trinidad) 51: 325–331.

Vitousek, P. M., and J. S. Denslow. 1986. Nitrogen and phosphorus availability in treefall gaps of a lowland tropical rainforest. *J. Ecol.* 74: 1167–1177.

Vitousek, P. M., and R. L. Sanford. 1986. Nutrient cycling in moist tropical forest. *Ann. Rev. Ecol. & Syst.* 17: 137–167.

Whitmore, T. C. 1984. *Tropical Rainforests of the Far East.* Second edition. Clarendon Press, Oxford.

Wyatt-Smith, J. 1963. Enrichment planting. In *Manual of Malayan Silviculture for Inland Forest.* Malayan Forest Records No. 23, Sections 6:1–6:15. Forest Department, Kuala Lumpur.

8 NURSERY AND VEGETATIVE PROPAGATION TECHNIQUES FOR GENETIC IMPROVEMENT AND LARGE-SCALE ENRICHMENT PLANTING OF DIPTEROCARPS

Pedro H. Moura-Costa

Much of the economy of Southeast Asian countries is derived from the logging of dipterocarp forests. In order to maintain the forest products industry and the economic returns of this sector, forest regeneration must be managed for sustainable yields. The high densities of natural forest stands in Sabah, East Malaysia (Newbery et al. 1992), allow timber extraction rates of up to 120 m³/ha[1] (Silam Forest Products, timber extraction figures). This rate of extraction, however, substantially disturbs the residual stand (Appanah and Weinland 1990; Nussbaum, Anderson, and Spencer, this volume). In some areas, the residual stocking and seedling bank of timber species are much reduced, and artificial regeneration must be used (Appanah and Weinland 1990).

Enrichment planting is a technique for promoting artificial regeneration in which seedlings of preferred timber trees are planted in the understory of existing forests and given preferential treatment to encourage their growth. Enrichment planting of dipterocarps has a long history (Appanah and Weinland 1993, and references therein), although in some cases survival rates of plants were low and growth rates disappointing (Awang and Sawal 1986; Chai 1975; Priasukmana 1989; Tang and Wadley 1976). Lack of plot maintenance during the initial years may contribute substantially to poor performance in these trials; most seedling damage and mortality occurs before plants reach 3 m in height (Wyatt-Smith 1963). Seedlings are particularly susceptible to damage by insect borers (Becker 1983; Daljeet-Singh 1975), browsing mammals (Becker 1985; Wyatt-Smith 1963), competing vegetation such as lianas and vines (Chai 1975; Priasukmana 1989; Tang and Wadley 1976), and physiological stress caused by drought (Awang and Sawal 1986) or excessive radiation (Sasaki and Mori 1981). Maintenance of plots is therefore essential during this initial phase. Subsequent silvicultural treatments such as Liberation Thinning (which involves the selective poison girdling of trees that impinge on the canopy of future timber trees) are also required to maintain growth rates (Fox and Chai 1982).

These requirements may be met only if a project receives long-term support. An example is the cooperative project between Innoprise Corporation Sdn. Bhd. (the commercial arm of Yayasan, Sabah) and the FACE Foundation (Forests Absorbing Carbon-dioxide Emission, the Netherlands), initiated in the Ulu Segama Forest Reserve (Sabah, Malaysia) with the objective of promoting the rehabilitation of logged forests to absorb CO_2 from the atmosphere (Moura-Costa 1993). The project involves large-scale enrichment planting of degraded forests using seedlings of dipterocarp trees. The project area comprises 25,000 ha of logged-over dipterocarp forests around the Danum Valley Field Centre, which also supports other research related to forest ecology and regeneration (see Marshall and Swaine 1992). The long-term nature of the Innoprise–FACE Foundation project (twenty-five years) allows the maintenance of plots, including four rounds of weeding a year for two years and follow-up silvicultural treatments.

NURSERY FACILITIES AND SOURCES OF PLANTING MATERIAL

Large-scale planting of dipterocarps is inherently difficult because planting material is scarce. Dipterocarps exhibit mast fruiting, with one to ten years between seeding years (Ashton, Givnish, and Appanah 1988), and their seeds are only briefly viable, preventing long-term storage (Sasaki 1980). In seeding years, large numbers of wild propagated seedlings (wildings) are found on the forest floor (Liew and Wong 1973) and can be transferred to the nursery. Wilding availability, however, depends on the occurrence of seeding years and is therefore unreliable.

Both seeds and wildings can be used as nursery stock. During the fruiting season, seeds must be collected daily (in order to avoid predation by insects or mammals) and then taken immediately to the nursery for germination. When fresh seeds are used, germination rates of up to 97% (n = 5,000, SE = 1.2) can be obtained after two weeks for *Dryobalanops lanceolata, Shorea leprosula,* and *Parashorea malaanonan.* When seeds are kept for more than two weeks, the germination rate is less than 50% (n = 5,000, SE = 4.2). If seeds are not collected from the forest, they quickly germinate, and the resulting wildings can be cultivated. Wildings are simply pulled by hand from the forest floor (one person can collect up to 180 wildings/hr), but this method often damages their root system. They must thus be given special care during an acclimatization period after transfer to the nursery. Wildings are watered and kept in plastic-covered chambers with high humidity, until they form a new root system. Survival rates are satisfactory with this system: 94.4% (n = 200, SE = 3.2) for *Shorea parvifolia* and

88.2% (n = 200, SE = 2.5) for *Dryobalanops lanceolata,* after a four-week acclimatization period.

Vegetative Propagation by Cuttings

Many researchers have investigated the use of vegetative propagation by cuttings to supply dipterocarp planting stock (Hallé and Hanif-Kamil 1981; Hamzah 1990; Kantarli 1993; Momose 1978; Smits 1983; Srivastava and Penguang Manggil 1981). Dipterocarps are considered difficult to root, however, and results are sometimes unsatisfactory. Techniques for vegetative propagation of dipterocarps have been developed in Danum Valley Field Centre (Moura-Costa and Lundoh, 1994a, 1994b) based on the methods used by Smits (1983), Leakey (1983), and Leakey, Chapman, and Longman (1982). These techniques involve taking two-node cuttings from the apical shoots of juvenile stockplants (either young seedlings or managed hedge orchards), trimming their leaves to approximately 30 cm^2, and dipping their basal ends in a fungicide solution. No auxins are used, because previous experiments showed that IBA, NAA, and 2,4–D (at 0.2, 0.8, and 3.0% w/w) suppress rooting of juvenile cuttings of some dipterocarps (Moura-Costa and Lundoh, 1994b). Cuttings are rooted in a mist propagator unit covered by a transparent plastic sheet (fig. 8.1)

FIGURE 8.1   Inside view of a closed-chamber mist propagator used in Danum Valley Field Centre, loaded with cuttings of different dipterocarp species.

FIGURE 8.2 A two-node cutting of *Dryobalanops lanceolata* beginning to form roots, after eight weeks in the propagation unit.

to increase relative humidity. After roots form (fig. 8.2), cuttings are potted and kept in plastic-covered chambers in the shade house for two weeks during the acclimatization stage, before exposure to normal nursery conditions. Mycorrhizal inoculation is done during potting, by adding soil containing inocula to the soil in the planting bags. Cuttings of *D. lanceolata* and several *Shorea* species have been produced using this method. Percentage rooting after twelve weeks is around 87% (n = 396, SE = 1.6) for *D. lanceolata* and 65% (n = 40, SE = 2.4) for *Shorea* spp.

Hedge orchards with stockplants are being established in Danum Valley Field Centre to guarantee a steady supply of planting material for cutting production. Seedlings are planted in lines in a partially shaded area adjacent to the research nursery and are managed to increase the number of orthotropic (upward-growing) shoots produced by each plant (fig. 8.3). This is done by removing the apical shoot and bending the stem, in order to break apical dominance and induce the sprouting of dormant stem nodes (Leakey 1983). Using this method, up to fifteen shoots can be formed per month by each stockplant.

This method of vegetative propagation is being experimentally modified to minimize the number of operations required for cutting production, thus reducing costs and enabling the method to be used for large-scale

FIGURE 8.3   Inside view of the large-scale nursery for raising dipterocarp plants. Note, in the back, the plastic-covered chambers used for acclimatization of wildings.

production of cuttings in our operational nursery. In an experiment in which cuttings of *Dryobalanops lanceolata* were set to root directly in 7×21 cm poly-bags containing forest topsoil, percentage rooting after twelve weeks in the mist unit was 83% ($n = 20$, $SE = 2.3$). Further trials are underway to test the rooting of cuttings in the plastic-covered chambers used for acclimatization of wildings, in order to determine if mist units are really needed.

Micropropagation

Tissue culture techniques for in vitro propagation of dipterocarps are also being investigated in collaboration with the Forest Research Centre (Sepilok, Sandakan, Malaysia). Linington's (1991) methods are used for shoot multiplication of dipterocarps. Experiments for rooting micropropagated dipterocarp plants used *Dipterocarpus intricatus* plantlets from cultures provided by I. Linington (Kew Gardens, England). When 0.8% IBA talcum formulations were applied to the basal end of the plantlets, 35% ($n = 20$) formed roots after twelve weeks in the mist-spray sand beds. This method could perhaps be improved to attain better percentages of rooting and establishment of plants ex vitro. Successful tissue culture systems would enable the large-scale rapid multiplication of clonal genotypes, with enormous advantages for commercial forestry (Bonga and Durzan 1987). A further improvement would be the study of somatic embryogenesis for artificial seed production (Moura-Costa, Viana, and Mantell 1993; Redenbaugh and Ruzin 1989), with potential for the mass propagation of tropical trees at a low cost (see Mascarenhas et al. 1988).

Clonal Propagation and Genetic Improvement of Dipterocarps

Previous trials of enrichment planting have used trees from natural populations, consisting of a wide range of genotypes with different growth rates. Vegetative propagation can be used to produce clones, reducing genetic variability of planting stock and permitting better-controlled studies of growth and development of dipterocarps in enrichment planting and the establishment of stands with more homogeneous growth rates. A further advantage of using vegetative propagation is the possibility of genetic improvement of planting stock. For example, threefold increases in biomass production of *Eucalyptus* have been achieved in Aracruz Celulose, Brazil, using selection and cloning (Campinhos and Ikemori 1983). To my knowledge, no work has been done on genetic improvement of dipterocarps; substantial gains are expected from selection and cloning of superior genotypes. The advantages and disadvantages to clonal populations of forest trees are discussed here.

Problems of Selection and Propagation of Adult Trees

Clonal forestry is typically based on the phenotypic selection of superior trees followed by vegetative propagation (Hartmann, Kester, and Davies 1990); however, selection based on mature trees has several problems. First, planting material from adult trees is physiologically mature and often more difficult to root, as Smits et al. (1990) have demonstrated for dipterocarps. The use of mature material for vegetative propagation might have been the reason for the low percentages of rooting achieved by Momose (1978) and Hallé and Hanif-Kamil (1981). All cuttings Momose prepared from mature tissues failed to root. In contrast, Smits (1983) and Hamzah (1990) observed high percentages of rooting when they used juvenile material. Therefore, mature material taken from adult trees needs to be rejuvenated in order to improve rooting ability (see Hackett 1985). Grafting mature scions onto juvenile rootstocks (Martin and Quillet 1974) or serial cuttings (Black 1972) are possible methods. An alternative is to induce the production of juvenile tissue from mature trees (Hackett 1985). Most dipterocarp trees do not coppice, however, and methods involving the manipulation of the whole tree, such as severe pruning (Mazalewsky and Hackett 1979), are not feasible.

Second, the selection of superior trees in the forest relies on phenotypic characteristics. The phenotypes of mature trees in natural forests, however, are strongly influenced by environmental factors, which can affect the expression of their genotypes (see Namkoong, Barnes, and Burley 1980). Furthermore, the recruitment of dipterocarp trees from the seedling bank in the forest is not necessarily linked to genotypic characteristics. According to the process of gap regeneration dynamics described by Whitmore (1984), dipterocarp seedlings stay in the understory for long periods and are liberated when gaps form in the canopy. The random process of gap formation benefits not the best genotypes but those that happen to be present in the gap. Therefore, selection of mature trees is not reliable and must be confirmed by carrying out clonal trials, which are thought to be impractical because of the slow growth of dipterocarps.

Selection at the Seedling Stage: Apical Dominance in Juvenile Material

Correlations between juvenile characteristics and desirable genetic traits of mature trees would allow the selection of plants at an early stage, circumventing some of the problems. Ladipo, Leakey, and Grace (1991) suggested the use of apical dominance of seedlings as a selection factor and found a positive correlation between this factor and height of *Triplochiton scleroxylon* trees after five years. The strong negative correlation between increase in total branch length and height increment observed in *D. lanceolata* seedlings (Moura-Costa, Viana, and Mantell 1993) suggests that this method may be applicable for dipterocarps. Strong apical dominance is also a desirable characteristic in the Ulu Segama Forest Reserve, where up to 40% of planted dipterocarp seedlings suffered damage to their apical meristems from browsing mammals or stemboring insects (L. Lundoh and M. Pinard, unpublished data, 1991). Plants with strong apical dominance produce a single replacement shoot, thus retaining a good stem form. Because apical dominance can be expressed at the seedling stage, selection of superior genotypes can be done in the nursery, which reduces the effect of environmental variability and provides juvenile material suitable for clonal propagation. When the predictive test for apical dominance (Ladipo, Leakey, and Grace 1991) was applied to seedlings of *Dryobalanops lanceolata,* the degree of apical dominance obtained after decapitation of 350 seedlings was highly variable, creating scope for selection (Moura-Costa, unpublished data, 1992). A proportion of the plants are being decapitated a second time to test whether this characteristic is consistently expressed, that is, genetically controlled. Further assessments are being carried out in order to determine how quickly plants regain apical dominance. Trials are also being conducted to test for a correlation between reestablishment of apical dominance and height increment of trees grown in the field.

The use of clonal material for forestry is controversial. Although timber yields can be substantially improved (Campinhos and Ikemori 1983), the risks of susceptibility to outbreaks of diseases or pests are usually higher in clonal stands than in sexually propagated populations. Therefore, selection intensity should be modest, allowing the use of a large number of clones to guarantee a diverse genetic base in the planted stands.

A second concern is the rooting system of plants derived from cuttings. Cutting-derived plants do not produce taproots, so they may be especially susceptible to windthrow or droughts. They are, however, able to produce a large root plate from which axillary sinker roots often develop (Leakey 1987). Furthermore, most dipterocarp trees do not form deep taproots (Baillie and Mamit 1983) but shallow lateral roots from which vertical sinkers extend. The same pattern of root systems can be expected for wildings. After wildings are collected, their taproot is often pruned to promote the development of adventitious roots. Although no old plantations of dipterocarps grown from cuttings exist, plots planted with dipterocarp wildings dating back to 1935 (see Tan et al. 1987) do not appear to suffer from any of the problems related to a poor root system.

CONCLUSIONS

Advances in vegetative propagation and nursery techniques have improved the reliability of supply and the quality of dipterocarp planting material. The development of efficient tissue culture techniques might provide enormous benefits for dipterocarp propagation, but further studies are still required to increase multiplication rates in vitro and percentages of rooting ex vitro. Whichever propagation system is used, costs of production must be kept low, because planting stock accounts for a substantial share of the costs of enrichment planting.

Genetically superior planting stock is urgently needed to improve growth rates of planted stands and increase the attractiveness of enrichment planting as a tool for the management of logged rainforests. If a correlation between apical dominance and growth rates of dipterocarp trees is confirmed, it can be used for a simple selection procedure. Clonal populations of fast-growing genotypes can be established in a short period, improving the quality of planted stands and reducing the time required for rehabilitation of logged forests by enrichment planting.

REFERENCES

Appanah, S., and G. Weinland. 1990. Will the management systems for hill dipterocarp forests stand up? *J. Trop. For. Sci.* 3: 149.

———. 1993. Planting quality timber trees in Peninsular Malaysia—a review. *Malaysian For. Rec.* No. 38. Forest Research Institute, Kepong.

Ashton, P. S., T. J. Givnish, and S. Appanah. 1988. Staggered flowering in the Dipterocarpaceae: New insights into floral induction and the evolution of mast fruiting in the aseasonal tropics. *Am. Nat.* 132: 44–66.

Awang, K., and P. Sawal. 1986. Initial performance of three indigenous species used in an enrichment planting of hill dipterocarp forest. In *Proceedings of the Workshop on Impact of Man's Activities on Upland Forest Ecosystem.* H. Yusuf, K. Awang, N. M. Majid, and S. Mohamed, editors. Universiti Pertanian Malaysia, Selangor.

Baillie, I. C., and J. D. Mamit. 1983. Observations on rooting in mixed dipterocarp forest, Central Sarawak. *Malay. For.* 46: 369–374.

Becker, P. 1983. Effects of insect herbivory and artificial defoliation on survival of *Shorea* seedlings. In *Tropical Rainforest Ecology and Management.* S. L. Sutton, T. C. Whitmore, and A. C. Chadwick, editors. Blackwell Scientific, Oxford.

———. 1985. Catastrophic mortality of *Shorea leprosula* juveniles in a small gap. *Malay. For.* 48: 263–265.

Black, D. K. 1972. The influence of shoot origin on the rooting of Douglas-fir stem cuttings. *Proc. Int. Plant Propagation Soc.* 22: 142–159.

Bonga, J. M. and D. J. Durzan, editors. 1987. *Cell and Tissue Culture in Forestry.* Martinus Nijhoff, Dordrecht, The Netherlands.

Campinhos, E., Jr., and Y. K. Ikemori. 1983. Mass production of *Eucalyptus* spp. by rooting cuttings. *Silvicultura* 8: 770–775.

Chai, D. N. P. 1975. Enrichment planting in Sabah. *Malay. For.* 38: 271–277.

Daljeet-Singh, K. 1975. A preliminary survey of insect attack on seedlings and saplings in Bukit Belata Forest Reserve. *Malay. For.* 38: 14–16.

Fox, J. E. D., and N. P. Chai. 1982. Refinement of the regenerating stand of a *Parashorea tomentella/Eusideroxylon zwageri* type of lowland dipterocarp forest in Sabah—a problem in silvicultural management. *Malay. For.* 45: 133–183.

Hackett, W. P. 1985. Juvenility, maturation and rejuvenation in woody plants. *Hort. Rev.* 7: 109–155.

Hallé, F., and Hanif-Kamil. 1981. Vegetative propagation of dipterocarps by stem cuttings and air-layering. *Malay. For.* 44: 314–318.

Hamzah, A. 1990. A note on the rooting of *Shorea bracteolata* stem cuttings. *J. Trop. For. Sci.* 3: 187–188.

Hartmann, H. T., D. E. Kester, and F. T. Davies. 1990. *Plant Propagation: Principles and Practices.* Fifth edition. Prentice-Hall International Editions, Upper Saddle River, N.J.

Kantarli, M. 1993. Vegetative propagation by cuttings in ASEAN region. Review Paper No. 1. ASEAN-Canada Forest Tree Seed Centre Project, Saraburi, Thailand.

Ladipo, D. O., R. R. B. Leakey, and J. Grace. 1991. Clonal variation in a four-year-old plantation of *Triplochiton scleroxylon* K. Schum. and its relation to the predictive test for branching habit. *Silvae Genetica* 40: 130–135.

Leakey, R. R. B. 1983. Stockplant factors affecting root initiation in cuttings of *Triplochiton scleroxylon* K. Schum., an indigenous hardwood of West Africa. *J. Hort. Sci.* 58: 277–290.

———. 1987. Clonal forestry in the tropics—A review of the developments, strategies and opportunities. *Commonw. For. Rev.* 66: 61–75.

Leakey, R. R. B., V. R. Chapman, and K. A. Longman. 1982. Physiological studies for tree improvement and conservation. Factors affecting root initiation in cuttings of *Triplochiton scleroxylon* K. Schum. *For. Ecol. & Mgmt.* 4: 53–66.

Liew, T. C., and F. O. Wong. 1973. Density, recruitment, mortality and growth of dipterocarp seedlings in virgin and logged-over forests in Sabah. *Malay. For.* 36: 3–15.

Linington, I. M. 1991. *In vitro* propagation of *Dipterocarpus alatus* and *Dipterocarpus intricatus*. *Plant Cell, Tissue & Organ Cult.* 27: 81–88.

Marshall, A. G., and M. D. Swaine, editors. 1992. Tropical rain forest: disturbance and recovery. Proceedings of a Royal Society Discussion Meeting. *Phil. Trans. Roy. Soc. Lon.* 335: 323–447.

Martin, B., and G. Quillet. 1974. Bouturage des arbres forestiers au Congo. Resultats des essais effectués Pointe-Noire de 1969 1973. Rajeunissement des arbres plus et constitution du parc bois. *Bois et forts des tropiques* 157: 21–40.

Mascarenhas, A. F., S. S. Khuspe, R. S. Nadgauda, P. K. Gupta, and B. M. Khan. 1988. Potential of cell culture in plantation forestry programs. In *Genetic Manipulation of Woody Plants*. J. W. Hanover and P. E. Keathley, editors. Plenum Press, New York.

Mazalewsky, R. L. and W. P. Hackett 1979. Cutting propagation of *Eucaliptus ficifolia* using cytokinin-induced basal trunk sprouts. *Proc. Intern. Plant Prop. Soc.* 29: 118–24.

Momose, Y. 1978. Vegetative propagation of Malaysian trees. *Malay. For.* 41: 219–223.

Moura-Costa, P.H. 1993. Large scale enrichment planting with dipterocarps, methods and preliminary results. In *Proceedings of the Yogyakarta Workshop: Bio-reforestation in Asia-Pacific Region*. K. Suzuki, S. Sakurai, and K. Ishii, editors. BIO-REFOR/IUFRO/SPDS.

Moura-Costa, P. H., and L. Lundoh. 1994a. A method for vegetative propagation of *Dryobalanops lanceolata* (Dipterocarpaceae) by cuttings. *J. Trop. For. Sci.* 6: 553–541.

———. 1994b. The effects of auxins (IBA, NAA and 2,4–D) on rooting of *Dryobalanops lanceolata* (kapur, Dipterocarpaceae) cuttings. *J. Trop. For. Sci.*

Moura-Costa, P. H., L. Lundoh, and S. C. Uren. In press. Effects of nitrogen and light on growth, development and nitrogen metabolism of *Dryobalanops lanceolata* Burck (Dipterocarpaceae) wildings grown in nursery. *J. Trop. For. Sci.*

Moura-Costa, P. H., A. M. Viana, and S. H. Mantell. 1993. *In vitro* plantlet regeneration of *Ocotea catharinensis*, an endangered Brazilian hardwood forest tree. *Plant Cell, Tissue & Organ Cult.* 35: 279–286.

Namkoong, G., R. D. Barnes, and J. Burley. 1980. A philosophy of breeding strategy for tropical forest trees. Tropical Forestry Papers No. 16. Commonwealth Forestry Institute, Oxford.

Newbery, D. McC., E. J. F. Campbell, Y. F. Lee, C. E. Ridsdale, and M. J. Still. 1992. Primary lowland dipterocarp forest at Danum Valley, Sabah, Malaysia: Structure, relative abundance and family composition. *Phil. Trans. Roy. Soc. Lon.* 335: 341–356.

Nussbaum, R., J. Anderson, and T. Spencer. This volume. Effects of selective logging on soil characteristics and growth of planted dipterocarp seedlings in Sabah.

Priasukmana, S. 1989. Planting experiments of dipterocarps in East Kalimantan. In *Fourth Round-Table Conference on Dipterocarps*. I. Soerianegara, S. S. Tjitrosomo, R. C. Umaly, and I. Umboh, editors. Biotrop Special Publication No. 41. Bogor, Indonesia.

Redenbaugh, K., and S. E. Ruzin. 1989. Artificial seed production and forestry. In *Applications of Biotechnology in Forestry and Horticulture*. V. Dhavan, editor. Plenum Publishing, New York.

Sasaki, S. 1980. Storage and germination of dipterocarp seeds. *Malay. For.* 43: 290–308.

Sasaki, S., and T. Mori. 1981. Growth responses of dipterocarp seedlings to light. *Malay. For.* 44: 319–345.

Smits, W. T. M. 1983. Vegetative propagation of *Shorea* cf. *obtusa* and *Agathis dammara* by means of leaf-cuttings and stem-cuttings. *Malay. For.* 46: 175–185.

Smits, W. T. M., I. Yasman, D. Leppe, and M. Noor. 1990. Summary of results concerning vegetative propagation of Dipterocarpaceae in Kalimantan, Indonesia. In *Breeding Tropical Trees: Population Structure and Genetic Improvement Strategies in Clonal and Seedling Forestry*. G. L. Gibson, A. R. Griffin, and A. C. Matheson, editors. Proceedings of the IUFRO Conference, Pattaya, Thailand. Oxford Forestry Institute, Oxford.

Srivastava, P. B. L., and Penguang Manggil. 1981. Vegetative propagation of some dipterocarps by cuttings. *Malay. For.* 44: 301–313.

Tan, S. S., R. Primack, O. K. Ernest, and H. S. Lee. 1987. The silviculture of dipterocarp trees in Sarawak, Malaysia, III: Plantation forest. *Malay. For.* 50: 148–116.

Tang, H. T., and H. Wadley. 1976. Report on the survival and development survey of areas reforested by line-planting in Selangor. FRI Res. Pamph. No. 67, Forest Research Institute, Kepong.

Whitmore, T. C. 1984. *Tropical Rain Forests of the Far East.* Clarendon Press, Oxford.

Wyatt-Smith, J. 1963. Enrichment planting. In *Manual of Malayan Silviculture for Inland Forest.* Malayan For. Rec. No. 23, Forest Department, Kuala Lumpur.

PART II Policy and Management

9 LONG-TERM ECOLOGICAL RESEARCH IN INDONESIA: ACHIEVING SUSTAINABLE FOREST MANAGEMENT

Herwasono Soedjito and Kuswata Kartawinata

The key to sustainable forest management is an understanding of forest dynamics. Such an understanding requires baseline data derived from long-term ecological studies of all types of natural forests. In Indonesia, few long-term ecological studies exist, though the rapid disappearance of the natural vegetation in many ecosystems makes such studies imperative. Because of the many immediate threats to Indonesian forests, conservation efforts must use the available information, though it is often incomplete.

In this chapter we describe Indonesian forests, review Indonesian ecological studies, and discuss human resources relevant to ecology and forest management. We also describe how these forests might be sustainably managed: enhanced conservation of biodiversity and natural forests must follow the promotion of sustainable use of forest resources.

INDONESIAN GEOGRAPHY AND VEGETATION

Indonesia is the world's largest archipelago, containing more than 17,000 islands stretching east-west for 5,200 km across the Sunda and Sahul continental shelves. Characterized by an enormously varied physical structure of high mountain ranges, volcanoes, alluvial plains, lakes, swamps, and shallow coastal waters, the archipelago exhibits a biological diversity and richness that is without comparison in Asia (FAO 1982–83; McNeely et al. 1990; Petocz 1989; Scott 1989; Whitten et al. 1987; Whitten, Mustafa, and Henderson 1987). The principal islands are Sumatra, Java, Kalimantan (Borneo), Sulawesi, and Irian Jaya (West Papua Island).

Indonesia's territory covers 7.7 million km², approximately 5.8 million km² (75.3%) of which is composed of marine and coastal waters. Indonesia bridges Indo-Malaya and Oceania, two of the earth's biogeographic realms. The Indo-Malayan region, located in the west, consists of Sumatra,

Kalimantan, Java, and Bali; the Australian region in the east consists of Sulawesi, Moluccas, the Eastern Sunda Islands, and Irian Jaya. The vegetation types to the east and west of the "Wallace Line" (Wallace 1876) are divided by a biogeographic boundary that extends from north to south along the Sunda Shelf. The natural vegetation on the shelf itself is principally Malesian, dominated by the commercially important tree family Dipterocarpaceae. The vegetation found to the east has greater affinities with the Austro-Pacific realm and is dominated by mixed tropical hardwood species. Deciduous monsoon forest occurs in seasonally dry areas, particularly in the southern and eastern islands, such as the Lesser Sundas and southern Irian Jaya. All together, the "outer" islands of Sumatra, Kalimantan, Sulawesi, Moluccas, and Irian Jaya contain approximately 10% of the world's tropical rainforests (Davies et al. 1986). With 114 million ha of closed forest, Indonesia possesses more tropical forest than any single African or Asian country and is second only to Brazil in tropical forest area worldwide (McNeely et al. 1990).

Indonesia also has extensive natural wetlands, including many of international importance. These wetlands are found in the low-lying alluvial plains and basins, flat-bottomed valleys, and mangrove estuaries of Sumatra, Kalimantan, and Irian Jaya (Haeruman 1988; IUCN 1992). Indonesia has at least forty-seven distinct natural ecosystems (Sastrapradja et al. 1989).

ECOLOGICAL STUDIES IN INDONESIA

The vegetation and ecology of Indonesian forests are not completely understood. Van Steenis (1935, 1957, 1972) gives a relatively complete, though somewhat outdated, qualitative description of the various vegetation types of Indonesia. Kartawinata (1978, 1989) further refines and applies this outline. Several general maps of Malesian vegetation (Direktorat Bina Program 1980) and detailed, large-scale vegetation maps of Sumatra (Laumonier, Gadrinab, and Purnajaya 1983; Laumonier, Purnajaya, and Setiabudi 1986; Laumonier, Purnajaya and Setiabudi 1987) are available. SEAMEO (South East Asian Minister Education Organization) BIOTROP produces a description of Indonesian vegetation (Laumonier 1992), and Kartawinata (1990) reviews natural vegetation studies in Indonesia. Whitmore (1984a, 1984b) summarizes the ecology of various types of rainforest ecosystems, but comprehensive studies of the ecology of Indonesia are relatively new, and more are urgently needed before the natural vegetation disappears. Many ecological studies in Indonesia are descriptive and focus mainly on the vegetation and dynamics of the lowland rainforests of Sumatra and Kalimantan (Kartawinata 1990; Laumonier 1992).

Indonesian ecosystems have received varied amounts of attention; mangrove ecosystems have been studied in some detail, but others have been investigated cursorily at best.

Limitations of Past Research and the Potential of Long-Term Studies

Most ecological studies in Indonesia have been descriptive, short term, and discontinuous. One descriptive study is a project on the vegetation of Sumatra. It aimed to demonstrate to managers the usefulness of vegetation in evaluating the potential of forest or agricultural land (Laumonier 1992) and was a joint venture between SEAMEO BIOTROP and Institut de la Carte Internationale de la Végétation, Toulouse, France. Examples of local and short-term ecological studies include government research projects. The National Biological Institute (formerly the Center for Research and Development in Biology) of the Indonesian Institute of Sciences was engaged in ecological research in Wanariset, East Kalimantan, to investigate the impact of human activities on plant communities in East Kalimantan (Kartawinata, Abdulhadi, and Partomihardjo 1981). The project was originally intended to develop permanent plots for comprehensive ecological research, but it was ended after five years because of financial constraints. Funding also limited the ecosystem study in Ketambe, Leuser National Park, Southeast Aceh, Sumatra. Another study, which focused on a 10.5–ha plot of lowland dipterocarp forest established in 1979 in Wanariset, East Kalimantan, was abandoned in 1983, and the results have not been published because of lack of facilities for analysis.

These studies were developed with a project-by-project approach, not a program approach. A single project that is not conceived as part of a long-term scheme to study forest dynamics is not likely to sustain the required funding and facilities (fig. 9.1). Indonesia should have large permanent plots for comprehensive ecological research and for use in comparative study with permanent plots established in other countries, such as those at Pasoh Forest Reserve in Peninsular Malaysia (LaFrankie 1992), Gunung Mulu National Park in Sarawak (Proctor et al. 1983), and La Selva Research Station in Costa Rica (Hartshorn 1983), or the half-century-old basic ecological research station at Barro Colorado Island in Panama (Hubbel and Foster 1986). As LaFrankie (1992) has stated, long-term and large-scale studies are urgently needed to manage Asia's tropical forests. Indonesia is the only country in Asia that still has considerable large tropical rainforest cover; these plots must be established.

In order to develop a long-term ecological research program, the parties must cooperate completely, and an appropriate approach must be estab-

FIGURE 9.1   Long-term studies of marked trees in permanent plots give information on the growth rates of trees. This information can be used to calculate the amount of wood that can be sustainably harvested from the forest. Shown here is a research team from the Sarawak Forest Department at Bako National Park. Photograph by R. Primack.

lished. The parties must be aware that they are establishing a multigenerational program, not a short-term project that lasts for the duration of one cabinet administration (five years) or that depends on the enthusiasm and motivation of a small group of individuals acting as project directors. Long-term ecological studies should be led by a solid program approach that is independent of the people who manage the project. The constraints on the previous examples developed because the projects' originators were not thinking in terms of permanent study plots and were not intending to develop programs that last for decades or even centuries. Yet such plots do exist in other countries—some examples are the Hubbard Brook Experimental Forest and the Hutcheson Memorial Forest Center (HMFC) at Rutgers University, both located in the United States (Botkin 1990; Pickett, Parker, and Fiedler 1992), and the Barro Colorado Island station in Panama (Hubbel and Foster 1986)—and similar plots would be possible in Indonesia. The usual duration of short-term plots is much shorter than one cycle of old-growth tropical rainforest succession (approximately fifty to eighty years). A program approach guarantees that permanent plots will remain intact for decades while ensuring that resources to study these plots are maintained for the long run.

The Status of Long-Term Research in Indonesia

Researchers from outside Indonesia have established long-term research projects at several sites in Kalimantan, focusing on wildlife ecology, phenology, and vegetation structure. Government officials have considered discontinuing these projects, however, because of difficult relationships or miscommunications with the foreign researchers. Specific problems are a lack of research reports and project goals that are not relevant to the conservation and management needs of the Indonesian government.

Other research programs do make important contributions to ecological research in Indonesia. One such program started in 1988, on the Joloi, Busang, and Murung tributaries in Barito Hulu, Central Kalimantan, as a collaborative venture between the University of Cambridge and the Ministry of Forestry and Center for Research and Development in Biology (Chivers 1992). This project intends to study the plant-animal interactions that contribute to the conservation of forested areas and to the regeneration of adjacent selectively logged areas. The Barito Ulu project has been successful in several respects: it has increased the number of ecological stations across the island of Borneo, and it combines elements of research, training, and forest management. Unfortunately, the project site is not located in a legally protected area, so there is no guarantee that the plots will exist long enough for adequate data collection, as forest conversion is very rapid in Central Kalimantan.

Some projects do not focus solely on forest dynamics research. A study that focuses on technical forest management is being developed in the Bukit Baka Bukit Raya National Park in west-central Kalimantan (Hendrison 1992). This area will be used as a field station for the Natural Resource Management Project (NRMP) and will be jointly managed by the National Planning Board (BAPPENAS) and the Ministry of Forestry; it is funded by the government of Indonesia, the International Tropical Timber Organization (ITTO), and the United States Agency for International Development (USAID) (Voss 1992). The NRMP is a cooperative policy and research program, scheduled to last five years (1991–1995); the purpose of the project is to improve natural resources management in Indonesia through (1) policy analysis of a broad range of issues and development of institutional capacity in natural resources policy analysis, and (2) field testing of improved policies and practices for the management of production forests and protected areas (BAPPENAS 1992). Though the research component is important, it is not the primary focus of the project.

Human Resources and Conservation

Technical forest managers must frequently consider the impact of management practices on local people. Studies that include interaction between people and forests have been established in the Kayan Mentarang Nature Reserve, East Kalimantan (Soedjito, Jessup, and Kartawinata 1992). This long-term program began in 1990 and is intended to establish the Apo Kayan-Mentarang Biosphere Reserve as a model for rural sustainable development compatible with conservation. The program was initiated by the World Wide Fund for Nature (WWF) and builds on a relation with the Center for Research and Development in Biology of the Indonesian Institute of Sciences and the Ministry of Forestry of Indonesia.

Development of human resources in the field of ecological studies is badly needed to achieve sustainable forest management. Indeed, basic knowledge of ecological issues is now becoming common worldwide, but few professional ecologists—no more than twenty-five—work in Indonesia. Because of this limited professional work force, 10% of the world's tropical forests cannot be understood and properly managed. Several crucial tasks face Indonesia. First, Indonesia needs to increase the number of trained personnel and make more efficient use of those already working in scientific and technological fields relevant to the conservation of biological diversity and the sustainable use of biological resources. Second, Indonesia must establish and maintain programs for scientific and technical education as well as training for managers and professionals in the fields of conservation biology and sustainable resource use. Third, Indonesia must promote and encourage understanding of the importance of monitoring data from permanent plot studies. This information is important for all levels of government, business enterprises, lending institutions, and educational programs. Without a dramatic increase in participation in conservation at all levels, Indonesia cannot hope to conserve its large and diverse forest resources.

STRATEGY AND OBJECTIVES OF LONG-TERM ECOLOGICAL RESEARCH

Long-term ecological research has four basic goals. First, the studies should document and monitor biological diversity, particularly plant diversity. Second, long-term research should provide continuing data on growth, mortality, regeneration, behavior, and dynamics of forest tree species. Third, these data should be used to establish a data base for research and education as a contribution to the conservation and management of many types of reserves and other protected areas in the world. Finally, the data accumulated in long-term programs should be used to develop models and methods for balancing human needs with biological diversity in the use of forest resources.

These goals will be achieved with the implementation of a wide range of activities that should enhance the accumulation of information and baseline data related to sustainable forest management. The strategy and activities will be as follows:

1. Conduct rapid assessment activities for providing a general view of biodiversity.
2. Survey entire target areas to create large-scale maps of ecosystems. These surveys will help identify and protect critical ecosystems that are legally unprotected but incorporated within a particular landscape system.
3. Select representative permanent plots for monitoring ecosystem changes.

The permanent-plot monitoring system provides a baseline distribution of tree species and describes habitats within a particular site at a particular time (Dallmeier 1992). If changes in these permanent plots are monitored, the impact of natural and artificial disturbances on species and community composition can be understood. This understanding can be used to predict future changes. Without adequate inventories, managers cannot measure change, identify gaps in knowledge, or set priorities to fill these gaps.

We believe that the strategy and objectives of long-term ecological research are consistent with the goals of a comprehensive conservation management program. To be effective, conservation of biological diversity should take into account *process* and *context* (Pickett, Parker, and Fiedler 1992). It is the *processes* that generate or maintain the species, communities, ecosystems, or landscapes, and the spatial and functional *context* that must be maintained.

The ongoing Kayan Mentarang Project in East Kalimantan is a good model. This project has been starting to establish field stations as well as permanent plots. The total area of this strict nature reserve (soon to become a national park) is 1.6 million ha, covering the headwaters of the Kayan and Mentarang rivers and their tributaries, in the far interior of East Kalimantan, bordering on the Malaysian states of Sarawak and Sabah (figs. 9.2 and 9.3). Elevations in the area range between 200 m and 2,500 m above sea level; the region has a wet tropical climate with annual rainfall around 4,000 mm. The major vegetation types are lowland mixed dipterocarp forest, lower and upper montane rainforest, heath forest (*kerangas*), and secondary forest near the villages. Conservationally important vertebrates, such as birds (about 250 spp., including 7 species of hornbills) and mammals (96 spp., including 8 species of

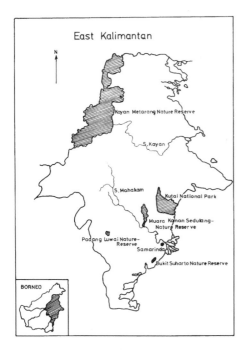

East Kalimantan

BORNEO

FIGURE 9.2   Nature reserves and national parks in East Kalimantan, Indonesian Borneo. The Kayan Mentarang Nature Reserve has an area of 1.6 million ha, covering headwaters of the Kayan and Mentanag rivers and their tributaries and bordering on the Malaysian states of Sarawak and Sabah.

FIGURE 9.3   The Nggeng River near the Lalut Birai research station is being used for the study of aquatic ecology.

primates), inhabit this unique area. It is also culturally diverse; at least twelve distinct ethnic-linguistic groups live in or near the reserve. The indigenous Dayak people, for example, have strong oral traditions, including significant forest-related knowledge and crafts. Historically, this area was the traditional homeland of several major cultural groups, and many archaeological remains have been found there.

This project is a collaboration of three agencies, each of which has assumed a different role: the Department of Forestry of Indonesia owns the area, the Indonesian Institute of Sciences acts as the principal scientific authority, and WWF is a nongovernment organization (NGO) that represents the reserve when it seeks research funding from outside sources. This project is expected to yield several important results. First, the program will establish biological inventories and a system for monitoring biological resources and human impact in conjunction with a long-term program for biological research. Second, plans are under way for integration of conservation in the reserve with regional economic development in cooperation with East Kalimantan's provincial planning authority (BAPPEDA). Third, a program of conservation education in the local communities, drawing on indigenous as well as scientific knowledge, will provide a broad understanding of conservation and encourage local people to pursue careers (as rangers, parataxonomists, field research assistants, and so on). Fourth, sustainable forest production will be established based on a zoning system that reflects both conservation priorities (in strict protection zones) and economic needs of local communities (in traditional-use zones). Fifth, local participation in conservation and resource management will be encouraged in the form of well-established, community-based organizations responsible for conservation management in their own areas. Finally, locally recruited, trained staff will carry out conservation and resource management activities in both strict protection and traditional-use zones.

CONCLUSION

Permanent ecological research stations where sufficiently large forest plots can be established should be built in Indonesia. Permanent plots for ecological studies differ from other natural areas because they are managed, but they can also be used as a control for other landscape studies as well as for basic studies on ecosystem functioning. In addition, permanent plots can be considered field laboratories where scientists can conduct experimental studies on the natural evolution of plant and animal communities and changes in structural and functional parameters of ecosystems resulting from human activity. By knowing about the process and context of natural systems, policymakers and managers will be able to make better choices in achieving the goal of sustainable forest management.

REFERENCES

BAPPENAS. 1992. *Joint Implementation Plan for the Natural Resources Management (NRM) and the Sustainable Forest Management (SFM) Projects 1991/1992 to 1994/1995.* BAPPENAS–Ministry of Forestry assisted by USAID and ITTO, Jakarta, Indonesia.

Botkin, D. B. 1990. *Discordant Harmonies.* Oxford University Press, New York.

Chivers, D. J. 1992. Project Barito Ulu—the role of animals in forest regeneration. In *Forest Biology and Conservation in Borneo.* G. Ismail, M. Mohamed, and S. Omar, editors. Center for Borneo Studies, Publication No. 2, Yayasan Sabah, Kota Kinabalu.

Dallmeier, F. 1992. Ecological long-term monitoring of protected areas and wildlands: Approaches to achieve the conservation of biological diversity. In *Additional Plenary Sessions and Symposium Papers of the Fourth World Congress on National Parks and Protected Areas.* February 10–21, 1992, Caracas, Venezuela, book 2. IUCN, Gland, Switzerland.

Davies, S. D., S. J. M. Droop, P. Gregerson, L. Henson, C. J. Leon, J. Lamlein Villa-Lobos, H. Synge, and J. Zantovka. 1986. *Plants in Danger: What Do We Know?* IUCN, Gland, Switzerland, and Cambridge, England.

Direktorat Bina Program. 1980. *Peta Tegakan Hutan Indonesia.* Direktorat Bina Program, Direktorat Jendral Kehutanan, Bogor, Indonesia.

FAO (Food and Agriculture Organization). 1982–83. *A National Conservation Plan for Indonesia.* 8 volumes. UNDP/FAO National Parks Development Project INS/78/061. FAO, Bogor, Indonesia.

Haeruman, H. J. 1988. Conservation in Indonesia. *Ambio* 17, 3: 218–222.

Hartshorn, G. S. 1983. Wildlands conservation in Central America. In *Tropical Rain Forest: Ecology and Management.* S. L. Sutton, T. C. Whitmore, and A. C. Chadwick, editors. Blackwell Scientific, Oxford, England.

Hendrison, J. 1992. Recommendations for controlled timber harvesting in the SBK forest concession. Natural Resources Management Project Report, USAID-Jakarta.

Hubbell, S. P., and R. B. Foster. 1986. Commonness and rarity in a Neotropical forest: Implications for tropical tree conservation. In *Conservation Biology.* M. Soulé, editor. Sinauer Press, Sunderland, Mass.

IUCN (International Union for the Conservation of Nature and Natural Resources). 1992. *Protected Areas of the World.* IUCN, Gland, Switzerland, and Cambridge, England.

Kartawinata, K. 1978. Ecological zones of Indonesia. In *Papers Presented at the Thirteenth Pacific Science Congress, Vancouver, Canada.* LIPI, Jakarta.

———. 1989. Progress on vegetation and ecological studies in Malesia with special reference to Indonesia. MAB Regional Seminar on Methods of Biological Inventory and Cartography for Ecological Management. October, Tokyo.

———. 1990. A review of natural vegetation studies in Malesia, with special reference to Indonesia. In *The Plant Diversity of Malesia.* P. Baas, K. Kalkman, and R. Geesink, editors. Kluwer Academic, The Netherlands.

Kartawinata, K., R. Abdulhadi, and T. Partomihardjo. 1981. Composition and structure of lowland dipterocarp forest at Wanariset, East Kalimantan. *Malay. For.* 44, 2 & 3: 397–406.

LaFrankie, J. V. 1992. Long-term and large-scale studies: Application to forest management in tropical Asia. In *Forest Biology and Conservation in Borneo.* G. Ismail, M. Mohamed, and S. Omar, editors. Center for Borneo Studies, Publication No. 2, Yayasan Sabah, Kota Kinabalu.

Laumonier, Y. 1992. The vegetation of Sumatra. In *Proceedings of the Workshop on Sumatra,*

Environment and Development: Its Past, Present and Future. BIOTROP Special Publication No. 46, SEAMEO-BIOTROP, Bogor, Indonesia.

Laumonier, Y., A. Gadrinab, and Purnajaya. 1983. *International Map of Vegetation and Environmental Conditions: Southern Sumatra, Scale 1:1,000,000.* Institut de la Carte Internationale du Tapis Végétal, Toulouse, France, and SEAMEO-BIOTROP, Bogor, Indonesia.

Laumonier, Y., Purnajaya, and Setiabudi. 1986. *International Map of Vegetation and Environmental Conditions: Central Sumatra, Scale 1:1,000.000.* Institut de la Carte Internationale du Tapis Végétal, Toulouse, France, and SEAMEO-BIOTROP, Bogor, Indonesia.

―――. 1987. *International Map of Vegetation and Environmental Conditions: Northern Sumatra, Scale 1:1,000,000.* Institut de la Carte Internationale du Tapis Végétal, Toulouse, France, and SEAMEO-BIOTROP, Bogor, Indonesia.

Leighton, M. 1990. A report on possible options for managing research and training in the Gunung Palung nature reserve. Report of the Gunung Palung Project in Rainforest Ecology, Management and Conservation, PPHA, Bogor, Indonesia.

McNeely, J. A., K. R. Miller, W. V. Reid, R. A. Mittermeier, and T. B. Werner. 1990. *Conserving the World's Biological Diversity.* IUCN, Gland, Switzerland, and Washington, D.C.

Petocz, R. G. 1989. *Conservation and Development in Irian Jaya: A Strategy for Rational Resource Utilization.* E. J. Brill, Leiden, The Netherlands.

Pickett, S. T. A., V. T. Parker, and P. L. Fiedler. 1992. The new paradigm in ecology: Implications for conservation biology above the species level. In *Conservation Biology.* P. L. Fiedler and S. K. Jain, editors. Chapman and Hall, New York and London.

Proctor, J., J. M. Anderson, P. Chai, and W. H. Wallak. 1983. Ecological studies in four contrasting lowland studies in Gunung Mulu National Park, Sarawak, I: Forest environment, structure and floristics. *J. Ecol.* 71: 237–260.

Sastrapradja, D. S., S. Adisoemarto, K. Kartawinata, S. Sastrapradja, and M. Rifai. 1989. *Keanekaragaman Hayati Untuk Kelangsungan Hidup Bangsa.* Pusat Penelitian dan Pengembangan Bioteknologi-LIPI, Bogor, Indonesia.

Scott, D. A., editor. 1989. *A Directory of Asian Wetlands.* IUCN, Gland, Switzerland, and Cambridge, England.

Soedjito, H., T. C. Jessup, and K. Kartawinata. 1992. The Apo Kayan-Mentarang Biosphere Reserve: A future conservation for development of Dayak Community in East Kalimantan, Indonesia. Paper presented at the Fourth World Congress on National Parks and Protected Areas, February 10–21, 1992, Caracas, Venezuela.

Van Steenis, C. G. G. J. 1935. Maleische Vegetatiescheten. *Tijdsch. Kon. Nederl. Aardrijksk. Genoot.* 52: 25–67, 171–203, 363–398.

―――. 1957. Outline of vegetation types in Indonesia and some adjacent regions. *Proc. Eighth Pacific Sci. Cong.* 4: 61–97.

―――. 1972. *The Mountain Flora of Java.* E. J. Brill, Leiden, The Netherlands.

Voss, R. 1992. Proposed Research Protocol for the Bukit Baka Research Station. Report of Research Station Consultant, BAPPENAS–Ministry of Forestry Assisted by USAID, Jakarta.

Wallace, A. R. 1876. *The Geographical Distribution of Animals.*

Whitmore, T. C. 1984a. *Tropical Rain Forests of the Far East.* Oxford University Press, Oxford.

―――. 1984b. A vegetation map of Malesia. *J. Biogeog.* 11: 461–471.

Whitten, A. J., S. J. Damanik, J. Anwar, and N. Hisyam. 1987. *The Ecology of Sumatra.* Gadjah Muda University Press, Yogyakarta, Indonesia.

Whitten, A. J., M. Mustafa, and G. S. Henderson. 1987. *The Ecology of Sulawesi.* Gudjah Muda University Press, Yogyakarta, Indonesia.

Eric Dinerstein, Eric D. Wikramanayake, and Mark Forney

The tropical moist forests of the Indo-Pacific region are among the oldest, most species-rich tropical forests on earth (Ashton 1969; Whitmore 1984). Primary forest cover has diminished rapidly, largely as a result of uncontrolled logging, conversion to agriculture, and burning (Collins, Sayers, and Whitmore 1991). Worldwide, conservation of tropical moist forests has become a critical issue, because although these forests compose only 6% of the earth's surface, they contain more than 50% of all species (Wilson 1989). Decline of biological diversity in the tropical zone is directly linked to deforestation.

Tropical moist forests in the Indo-Pacific region are a composite of at least nine major vegetation types that vary in richness, endemism, degree of deforestation, and commercial value. Strategies designed to conserve these forest types must integrate biological variables with threat factors using predictive, quantifiable methods. In 1993 we presented a new approach to setting priorities for investments in biodiversity conservation: a Conservation Potential/Threat Index (CPTI) (Dinerstein and Wikramanayake 1993). The CPTI forecasts how deforestation during the 1990s will affect conservation of forested habitats and the potential for establishment of new forest reserves. The index compares biological richness with size of protected areas, remaining forest cover, and severity of deforestation at various geographic scales, emphasizing habitat rather than species-oriented conservation.

In this chapter we adapt the CPTI to assess the conservation status of tropical moist forests in the Indo-Pacific region. We begin by identifying the main reservoirs of tropical moist forest and lowland rainforest from a regional perspective and forecast rates of loss among countries. We place particular emphasis on lowland rainforest, because it is perhaps the most species-rich vegetation type as well as the most threatened. Lowland rainforests are generally accessible, contain valuable timber, and are prime locations for establishing oil palm and rubber plantations and other crops. Second, we assess forest resources inside and outside protected areas and analyze the number and area of tropical moist forest reserves by country and biogeographic unit. Third, we determine conservation priorities at the national scale based on national deforestation rates, species richness and endemism, and present commitment to conservation in the form of existing protected areas, the number of large protected areas, and funds dedicated to conservation. Fourth, we assess the spatial distribution of protected areas to determine clustering and the extent to which smaller remnants of protected forest can be integrated into a larger conservation matrix. We also identify where large reserves can be established at a regional or biounit scale that transcend national boundaries to contribute to a representative network of tropical moist forest reserves in the region. We then recommend a portfolio of investments designed to more effectively conserve this dwindling global resource.

ANALYSIS OF THREATS TO TROPICAL FORESTS USING
CONSERVATION POTENTIAL/THREAT INDEXES

For the purposes of this discussion, tropical moist forest cover includes the forest types listed in IUCN 1986a: lowland rainforest, heath forest, mangroves, peatswamp forest, forests on limestone, forests on ultrabasics, tropical semievergreen forest, tropical montane evergreen forest, and monsoon forest. Subtropical moist deciduous forest, prevalent over much of India, southern Nepal, Bhutan, and several other countries, is not considered. To standardize the analysis, we use percentage of remaining tropical moist forest cover rather than absolute forest cover to rank countries regardless of their size. The projected amount of tropical moist forest habitat remaining in ten years' time is calculated by subtracting the amount of forest lost during the next ten years from the existing forested area, given a constant rate of deforestation (appendix 10.1). This estimate is probably conservative because annual deforestation rates in many countries are increasing (WRI 1990). We assume that park coverage (expressed as percentage of country under formal protection) will not expand beyond extensions proposed as of 1991 and that deforestation rates will remain constant during the next ten years.

We use rates of deforestation, rather than rates of degradation or fragmentation, to estimate threats to biodiversity because data for the latter two variables are unavailable for most countries. However, they are likely to be strongly correlated with rates of deforestation, a partial test of which is provided in this chapter. Deforestation rates are calculated for entire countries rather than for administrative or biogeographic units, which limit interpretation of threats. Deforestation is a more serious problem for South Asian countries and China, where tropical moist forest cover is minimal, than for Southeast Asian countries, where tropical moist forest accounts for much of the remaining forest cover. In South Asian nations, areas rich in tropical timbers may have higher deforestation rates than subtropical and temperate forested areas. Estimates of remaining habitat in several countries differentiate between primary and degraded forest cover, and where such data are available, only primary forest cover is used.

To position each country along the vertical axis of the CPTI, we assume that all proposed expansions of existing reserves and establishment of proposed reserves will occur by the end of the 1990s. This assumption is probably unrealistic, but it provides a glimpse of the potential for conservation that could be realized with major international support and local commitment. A number of protected areas exist only on paper and are not truly protective, whereas others that are operational lack proper management and are being degraded over time. Our threat index, at present, does not convey the effectiveness of park protection. Thus, the predictions of our index must be viewed as the most optimistic scenario possible for many countries in the Indo-Pacific region.

Assessing Biological Richness of Tropical Moist Forest Countries

We focus our analysis at the species level and assume that all taxonomic groups have equal importance. To assess species richness, we have tabulated the number of species per country for the following taxonomic groups: mammals, birds, reptiles, amphibians, freshwater fishes, swallow-tail butterflies, and vascular plants (see Dinerstein and Wikramanayake 1993, app. A). Countries are ranked by species richness and endemic species. For the three largest countries that contain limited tropical moist forest—India, China, and Australia—we edited country lists of species to include only tropical species (data on fauna for China and India are incomplete).

Invertebrates are greatly underrepresented, so our data base underestimates species richness and endemism for countries covered by extensive tracts of tropical moist forests known to be rich in invertebrates. Endemism is indexed by the number of endemic mammals, birds, and vascular plants per country. Information on other taxa is unavailable for all countries. Data on species richness and endemism by country and for Indonesian, Malaysian, and Philippine biogeographic units used to construct figures here are available in Dinerstein and Wikramanayake 1993, apps. A and B.

Landscape Analyses

We use forest cover maps published by Collins et al. (1991) and maps available for most countries in Geographical Information System (GIS) format to determine the degree of clustering and connectivity among protected areas containing tropical moist forest and similar, adjacent unprotected areas. For this analysis, we consider reserves to be nearest neighbors if distances are less than 50 km and the intervening habitat is not indicated to be degraded forest or non-forested habitat. We also count the number of reserves contiguous with forested areas of less than 1,000 km². We ignore national boundaries in these analyses, and we consider protected areas to be isolated from one another if separated by large bodies of water or mountain ranges or located on islands.

Abramovitz (1991) and the Global Environment Facility (GEF) (UNDP, World Bank, and UNEP 1991) provide data on the amount of funds invested by U.S. institutions during fiscal year 1989 in biodiversity conservation in each country. We express investment as dollars invested per km² of remaining tropical moist forest habitat and by size of protected forests within each country. We have not been able to determine how extensively internal and external funds are earmarked for tropical moist forest reserves. For those countries covered mostly by forest types other than tropical moist forests, budgets for tropical forest conservation have been estimated by the product of the total budget and the fraction of the area under protection that contains tropical moist forest.

REGIONAL PATTERNS

Indonesia is by far the largest reservoir of tropical moist forest in the Indo-Pacific region (fig. 10.1), containing 42% of the regional total and nearly half of the remaining lowland rainforest, though it occupies just 7% of the land area. Papua New Guinea, Burma, and Malaysia are also prominent in their contribution in spite of the relatively small total land area. Among the three largest countries, only India contains much tropical moist forest, but it is highly concentrated and only a small amount of lowland rainforest remains.

The impact of recent deforestation is evident in that five of the nineteen countries maintain less than 20% of their country area under tropical moist forest (fig. 10.2). Several other countries will reach this threshold within fifty years, assuming constant rates of deforestation. Only Fiji, Papua New Guinea, and the Solomon Islands are likely to maintain significant forest cover, based on current rates of deforestation. Australia and China are not included in this assessment because tropical moist forests have never exceeded 20% of the land area.

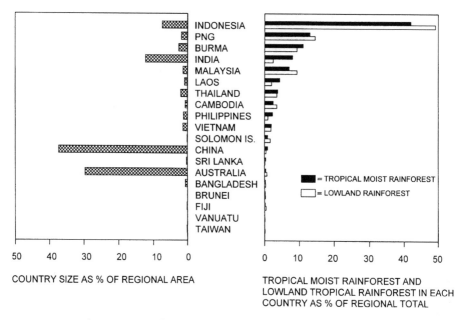

FIGURE 10.1 A comparison of tropical moist forest and lowland rainforest per country with country size expressed as a percentage of the regional total.

The legal status of protection and the absolute amount of remaining forest varies widely in the region (fig. 10.3). Indonesia's role as a regional reservoir is apparent, as is remnant forest cover in most other countries. Assuming that proposed extensions are achieved, five of the countries containing the largest reservoirs of tropical moist forest will be able to conserve at least 20% of the remaining forests in protected areas, with India on the boundary of this category. Taiwan, Brunei, Sri Lanka, and Australia have most of their remaining tropical moist forest under formal protection but contribute little to regional coverage. Burma and Papua New Guinea have extensive areas remaining, but little is under formal protection.

The potential for conservation and the threat to remaining tropical moist forests outside protected areas over the next decade is forecast in the CPTI (fig. 10.4). Assuming all proposed extensions of protected area systems are gazetted, only five countries will have both large amounts of tropical moist forest under protection and large areas remaining outside protected areas at the end of the 1990s (category I). Cambodia is included in category I at present, but recent reports of extensive clearing of forests may reduce it to category II. Thailand and Sri Lanka will have little tropical moist forest remaining outside protected areas (category II). High potential for increasing the conservation

Dinerstein, Wikramanayake, and Forney

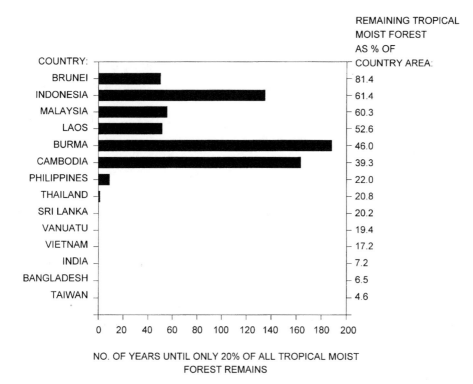

REMAINING TROPICAL MOIST FOREST AS % OF COUNTRY AREA:

COUNTRY:	
BRUNEI	81.4
INDONESIA	61.4
MALAYSIA	60.3
LAOS	52.6
BURMA	46.0
CAMBODIA	39.3
PHILIPPINES	22.0
THAILAND	20.8
SRI LANKA	20.2
VANUATU	19.4
VIETNAM	17.2
INDIA	7.2
BANGLADESH	6.5
TAIWAN	4.6

NO. OF YEARS UNTIL ONLY 20% OF ALL TROPICAL MOIST FOREST REMAINS

FIGURE 10.2   Threats to tropical moist forests by country, expressed in the number of years remaining until only 20% of the country's area is forested.

and sustainable use of tropical moist forests exists for Papua New Guinea, the Solomon Islands, and Burma (category III). The remaining countries examined have critically depleted their tropical moist forests and in the 1990s must protect what little remains (category IV).

At the national level, the biologically rich countries of Indonesia, Malaysia, and the Philippines show high internal variability in status of protection and size of remaining unprotected forest areas (fig. 10.5). Most Indonesian units have a high potential for conservation, Malaysian units are intermediate, and Philippine units are under great threat. The use of 10% protected and 40% forest cover as additional criteria may seem unrealistic for the Philippines units and Java and Bali, but several subunits of Indonesia, and Indonesia itself, already achieve or approach these targets. Use of the 10% line of protection is justified in that most of the fourteen biogeographic units considered are richer in species and endemics and higher in beta diversity than some entire countries in the Indo-Pacific region. Indonesia shows perhaps the greatest variation, spanning Java and Bali (3% forest cover, category II) and Irian Jaya (77% forest cover,

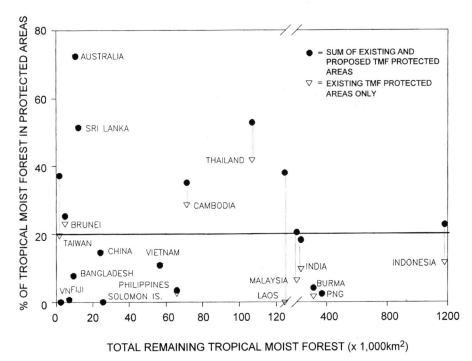

FIGURE 10.3   Status of protection of remaining tropical moist forest in Indo-Pacific countries. The percentage of tropical moist forest in protected areas is expressed as a percentage of the remaining forest cover. *VN* = Vanuatu. Note the shift in scale after the break along the *x*-axis.

category I). The placement of Sumatra in category I is misleading in that low-land rainforest habitat is under great pressure and underrepresented in terms of coverage; most large reserves are lower montane and montane forests. Significant areas of Irian Jaya, Maluku, Kalimantan, and Sulawesi are designated for protection, and adequate forest remains (category I).

Nearly twice as much land area is protected in Peninsular Malaysia and Sabah than in Sarawak (see fig. 10.5). Even though Sarawak contains considerably more forested area than the other two states, it is under great threat, and many important areas have been logged. The crisis in the Philippines is illustrated in the way all subunits, with the exception of Palawan, cluster in or near category IV. Palawan (category III) has the highest percentage of remaining forest cover; Mindanao and Luzon have larger areas of forest cover (in km²), but only a small amount is under formal protection. Data on the ratio of primary versus secondary and selectively logged forest, when available, will refine the priorities identified by a subnational CPTI.

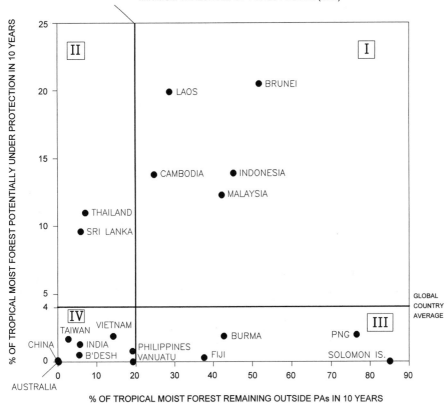

FIGURE 10.4 Conservation potential/threat index (CPTI). Unprotected forest remaining at the end of the 1990s is calculated using present forested areas and current deforestation rates. The variables graphed on each axis are presented as a percentage of country area. The predictions of the CPTI assume that all proposed protected areas are established and proposed expansions of existing reserves are completed by the end of the 1990s. Note the extensive areas slated for protection in Laos and Cambodia, none of which were truly operational as of 1993. *Category I:* Countries with a relatively large percentage (> 4%) of forests under formal protection and with a high (> 20%) proportion of unprotected forested areas still left in 2000. *Category II:* Countries with a relatively large percentage of forests (> 4%) under formal protection but with little (< 20%) unprotected forests left in 2000. *Category III:* Countries with a relatively low percentage (< 4%) of forests presently protected. Under current deforestation rates, however, these countries will still have a relatively large (> 20%) proportion of their unprotected forests remaining in 2000. *Category IV:* Countries with a relatively low proportion (< 4%) of forests presently protected. Under current deforestation rates, in 2000 little unprotected forest will remain.

BIOLOGICAL RICHNESS AND REMAINING HABITAT

Indexes of species richness and endemism, arranged by country and by category, reveal several general patterns. The two most species-rich countries, Indonesia and Malaysia, are also the most important reservoirs of tropical moist forest (fig.

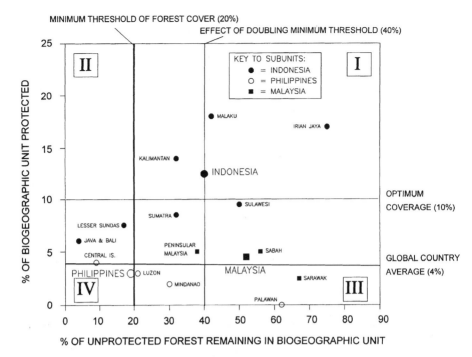

FIGURE 10.5   Application of the CPTI to biogeographic and administrative units for Indonesia, Malaysia, and the Philippines. The figure illustrates the effect of doubling the minimum threshold of forest cover outside protected areas and increasing protected habitat to optimum levels.

10.6a; see also appendix 10.1 and fig. 10.1). The most species-rich countries with the least forest cover are the Philippines and Vietnam. With the exception of Papua New Guinea, species richness in South Pacific countries is moderate to low relative to mainland countries. Levels of endemism reflect geographical isolation, being highest in such island nations as Indonesia, Papua New Guinea, Philippines, Sri Lanka, and Fiji.

Malaysian administrative units are richest in bird and mammal species, poor in endemics, and intermediate in terms of remaining forest cover (fig. 10.6b; see also fig. 10.5). Peninsular Malaysia has more bird and mammal species, but endemism is greater in Sabah and Sarawak than in Peninsular Malaysia. Inclusion of data on plant endemism and richness would clearly illustrate the conservation importance of Sabah and Sarawak (P. Ashton, personal communication, 1992). Indonesian biogeographic units vary considerably in the number of bird and mammal species, are relatively rich in endemics, and will retain extensive forest cover in most units (category I). In the Philippines, Palawan is low in endemism and species richness relative to the four other units;

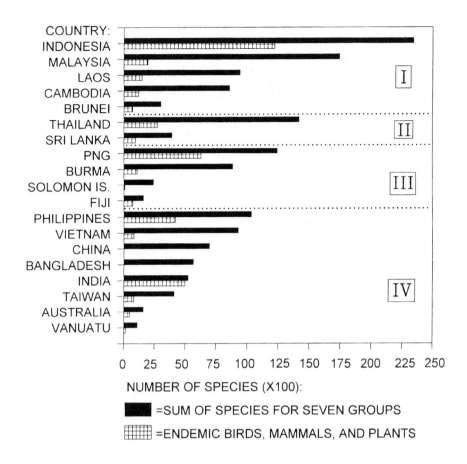

COUNTRY:

FIGURE 10.6a Indexes of species richness and endemism by category and by country as designated in fig. 10.4.

biogeographically, Palawan is more similar to Borneo than to other areas in the Philippines. The Central Islands constitute the only unit in category IV; the limited amount of remaining forest cover in this unit suggests that, of the units richest in endemic birds and mammals, it faces the most severe threats from habitat loss.

Biogeographic Patterns

The regional distribution of tropical moist forests by biogeographic unit (hereafter biounits) and the level of protection in each biounit show wide regional variation (fig. 10.7a and appendix 10.2). Twenty of the twenty-five biounits do not exceed 5% of the regional total (categories II and IV). Assuming all

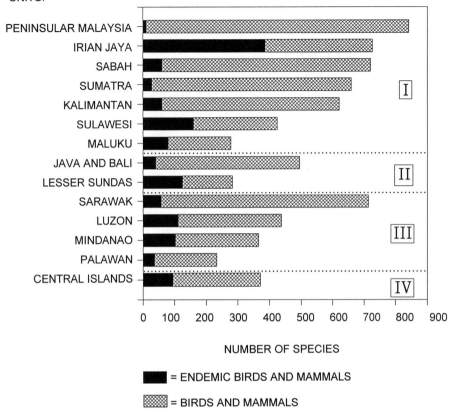

FIGURE 10.6b Bird and mammal species richness by category and by administrative or biogeo-graphic unit as designated in fig. 10.5. Note that the bars representing bird and mammal species and endemism both begin at zero.

proposed areas are established, only ten biounits would have more than 20% of remaining tropical moist forests in protected areas (categories I and II). All five of the biounits in categories I and III are extremely rich in species and endemics, clearly indicating a regional investment priority.

The regional distribution of lowland rainforest by biogeographic unit reveals that most of this forest type is concentrated in only four of the nineteen biounits (fig. 10.7b and appendix 10.2), none of which conserves more than 20% of this resource in existing reserves (category III). Less than 5% of the regional total of lowland rainforest is found in fifteen of the nineteen biounits, and adequate protection (more than 20%) is currently afforded in only three of

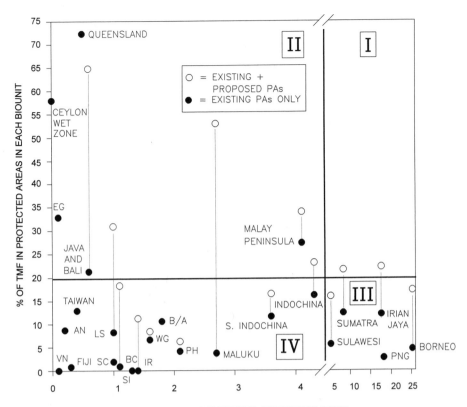

FIGURE 10.7a The conservation status of tropical moist forests and lowland rainforests from a biogeographic perspective. A comparison of the percentage of tropical moist forest in each biogeographic unit as a percentage of the regional total vs. percentage protected in each biounit.

these biounits (category II). The small size of some biounits and the limited absolute amount of tropical moist forest or lowland rainforest will limit the biounits that can place 20% of the remaining lowland forests in protected areas (such as Taiwan, Fiji, and South China).

Size Distribution of Individual Protected Areas by Country and Biogeographic Unit

Almost 46% of existing and proposed protected areas containing tropical moist forest are less than 300 km², and 70% are less than 1,000 km² (fig. 10.8); of reserves that contain at least 20% lowland rainforest, 46% are less than 300 km² and 74% are less than 1,000 km². The mean size of existing and proposed protected areas containing tropical moist forest is 644 km² (n = 848, SD = 1550) ver-

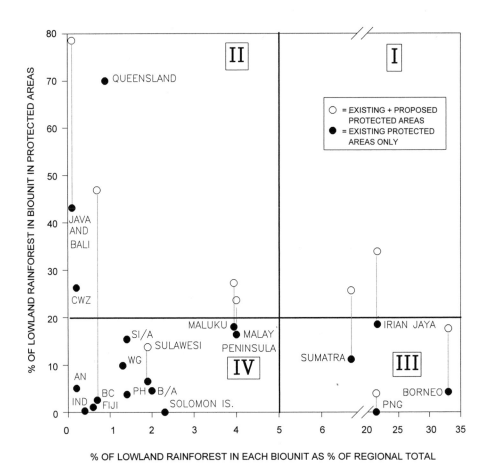

FIGURE 10.7b A comparison of the percentage of lowland rainforest in each biogeographic unit as a percentage of the regional total vs. percentage protected in each biounit. *AN* = Andaman Islands; *B/A* = Bengal/Assam; *BC* = Burmese coast; *EG* = Eastern Ghats; *IND* = Indochina; *IR* = Irrawaddy; *LS* = Lesser Sundas; *PH* = Philippines; *PNG* = Papua New Guinea; *SC* = South China; *SI* or *SI/A* = South Indochina and Annam; *VN* = Vanuatu; *WG* = Western Ghats.

sus 587 km² for nontropical forest reserves (*n* = 747, *SD* = 2061). Of an additional 122 reserves with unknown official status, 70% were less than 100 km².

Indonesia, Malaysia, and Thailand contain 69% (*n* = 92) of the existing and proposed large tropical moist forest reserves (greater than 1,000 km²), with 51% of the region's tropical moist forests located in these units (fig. 10.9 top and appendix 10.1). Overall, 67% of the region's tropical moist forests and 78% of its lowland rainforests are in large reserves. Six of the nineteen countries lack large reserves, and four contain only one large reserve. Indonesia's proposed and existing large reserves alone contain 88% of the protected lowland rainforest.

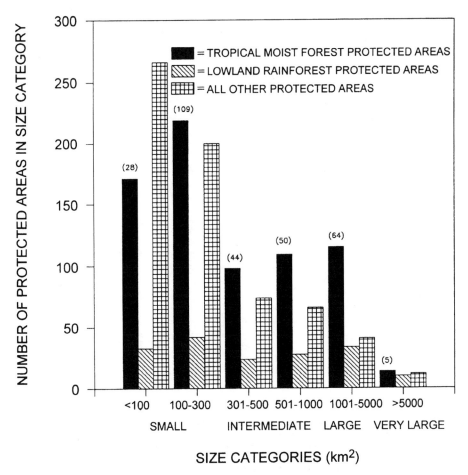

FIGURE 10.8 Size distribution of protected areas containing tropical moist forest and lowland rainforest in the nineteen countries considered in the region versus all other reserves in twenty-three countries of the region (excluding Australian protected areas outside of Queensland). The number of protected areas in each category includes existing and proposed reserves; in parentheses above the black bars is the number of proposed reserves.

Only thirty-five large reserves encompass elevational gradients ranging from lowland to montane tropical forest (fig. 10.9 bottom).

Most of the large tropical moist forest and lowland rainforest reserves are located in biogeographic units restricted to or overlapping with Indonesian territory (such as Borneo) (fig. 10.10). Central Indochina has a number of large reserves but none that contain at least 20% lowland rainforest. Existing reserves that span lowland and montane forests are most common in Borneo, Central Indochina, Sumatra, and Sulawesi, but Irian Jaya reserves con-

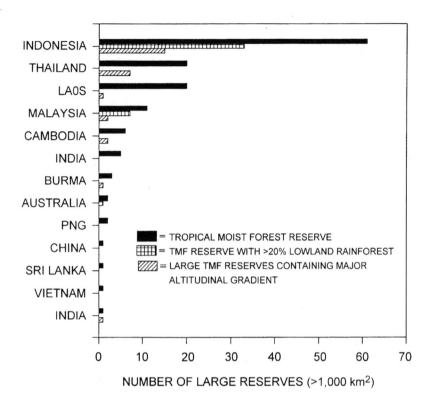

NUMBER OF LARGE RESERVES (>1,000 km²)

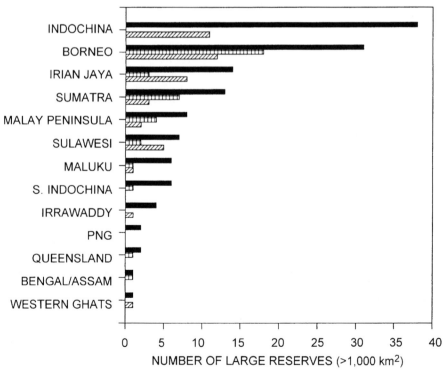

NUMBER OF LARGE RESERVES (>1,000 km²)

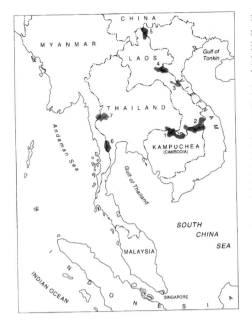

FIGURE 10.10 Potential transfrontier reserves for Indochina. Reserve 1 (15,489 km²) includes Preah Vihea, Cambodia (14,670 km²), Phnom Dongrak, Thailand (316 km²), Yot Dom, Thailand (203 km²), and Dang Kanh Thuong, Laos (300 km²). Reserve 2 (4,550 km²) includes Hondrai Sou, Cambodia (1,200 km²), Xe Piane, Laos (700 km²), Attapeu/Nam Khong, Laos (1,000 km²), Xe Kaman, Laos (1,200 km²), and Mom Rai Ngoc Vin, Vietnam (450 km²). Reserve 3 (860 km²) includes Nam Theun/On, Laos (700 km²), and Vu Quang, Vietnam (160 km²). Reserve 4 (more than 2,000 km²) includes Ahn Son, Vietnam (no data), and Lai Leng, Laos (2,000 km²). Reserve 5 (2,620 km²) includes Phou Dene Dinh, Laos (900 km²), and Muong Nhe, Vietnam (1,760 km²). Reserve 6 (3,404 km²) includes Kraeng Krachan NP, Thailand (2,915 km²), Mae Nam Prachi WS, Thailand (489 km²), and forested areas of Burma (about 2,000 km²). Reserve 7 (6,222 km²) includes Thung Yai Naresuan WS, Thailand (3,647 km²), Huay Kha Khaeng WS, Thailand (2,575 km²), and forested areas of Burma (about 750 km²).

tain the widest elevational ranges and, along with Papua New Guinea, have the greatest potential for creating similar types of reserves.

LANDSCAPE PATTERNS: SPATIAL DISTRIBUTION OF PROTECTED AREAS

Although protected areas in the region are small, the landscape level CPTI reveals that most countries and many biounits have the potential to manage reserves within larger forested landscapes (figs. 10.11 top, 10.11 bottom, and appendix 10.3). Analysis of spatial distribution of protected areas indicates that reserves in Sarawak, Brunei, Kalimantan, Irian Jaya, Sabah, Sumatra, Sulawesi, the Western Ghats in India, southern Laos, Thailand, Cambodia, Vietnam, and

FIGURE 10.9 Distribution of large reserves (> 1,000 km²) (*top*) by country and (*bottom*) by biogeographic unit in the Indo-Pacific region. Only lowland rainforest reserves containing more than 20% of this forest type are considered. Reserves with major elevational gradients range from lowland rainforest to montane forest at 1500 m or higher. (For a breakdown of existing versus proposed reserves, see appendixes 10.1 and 10.2). Note that all reserves in Laos are proposed; the contribution of Laos boosts the Indochina total for large reserves above Indonesian subunits.

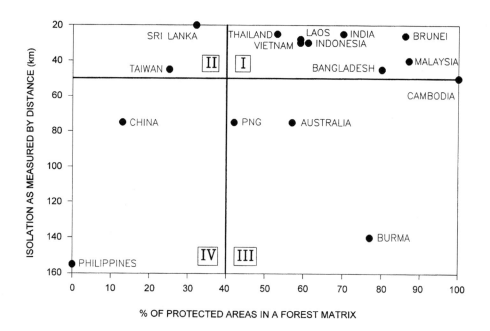

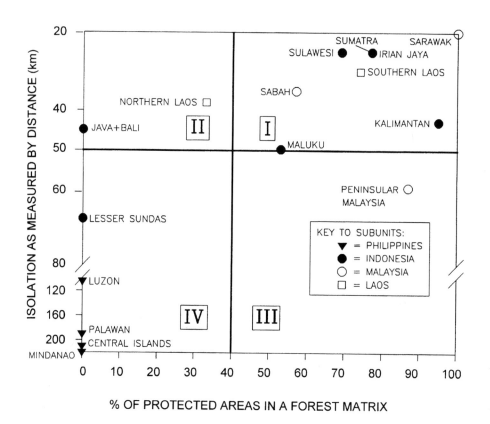

Dinerstein, Wikramanayake, and Forney

Bangladesh are potentially the units most easily integrated into larger forest units for conservation management; moreover, protected areas are on average relatively clustered. Protected areas in the Lesser Sundas, Java, Bali, southern China, Taiwan, and all the units in the Philippines will remain isolated fragments. The high within-country variation of relative isolation of protected areas is best illustrated by Indonesian and Laotian units (see fig. 10.11 bottom). Indonesia spans the isolated units of the Lesser Sundas, Java, and Bali and the more integrated units of Kalimantan, Irian Jaya, and Sumatra. Southern Laos, a regional reservoir of tropical moist forest, contains a number of proposed reserves with potential for landscape management approaches, as opposed to northern Laos, which occupies a separate biounit and is deforested and degraded.

Many of the countries occupy the same categories in the CPTI, as shown by examining forest cover at the national scale (see figs. 10.4 and 10.5) as at the landscape scale (see figs. 10.11 top and 10.11 bottom). This suggests a strong correlation between deforestation and fragmentation of forests. There are, however, notable exceptions: Bangladesh and Vietnam have suffered severe deforestation, but the remaining tropical moist forest reserves are relatively clustered and have high potential to be incorporated into larger conservation units. These figures require additional data on the status of adjacent forests to incorporate areas slated for logging, already degraded but not indicated as such on current maps, or other changes in land use that reduce potential for integrated forest management.

The difficulties in creating large reserves within countries, particularly those with little remaining forest cover, can be partially overcome by creating transfrontier, multinational reserves where forest cover is contiguous across national boundaries. The countries that comprise the region of Indochina are a good example of the potential for transfrontier reserves because forests in these countries are biologically rich, clustered near borders, and often under less in-

FIGURE 10.11   CPTI analysis of spatial distribution of existing and proposed protected areas containing tropical moist forest (*top*) by country and (*bottom*) by selected biogeographic or administrative units. Note that the scale of the *y*-axis is reversed to keep the position of the four categories consistent in all figures. Countries or biounits in category I have a high percentage of reserves clustered near other reserves and a high percentage situated either within or attached to a larger forest matrix that exceeds 1,000 km². Most remaining reserves of countries or biounits in category II are clustered, but the countries have little opportunity to integrate reserve management within a larger forested landscape. Reserves in category III countries or biounits are more isolated from one another but could be managed effectively as a larger landscape unit. Category IV countries and biounits contain many isolated reserves with little potential for management within a larger forest matrix.

Conserving Tropical Moist Forest

tense pressure from logging interests than forests elsewhere in each country (see fig. 10.10). In each case where we have identified a potential transfrontier reserve, forest on at least one side of the border is under some form of formal protection. Heavy exploitation of possible sites in Cambodia by the Khmer Rouge and by Thai logging companies, much of it illegal, reduces the potential for transfrontier reserves and should be halted. Other areas with high potential for transfrontier reserves include Borneo and New Guinea.

With the exception of Australia, Brunei, Malaysia, and perhaps Thailand, tropical moist forest in the region is concentrated in countries least able to adequately finance management and extension of the existing protected areas system (table 10.1). The high sum invested per km² in the Philippines by the GEF serves as a warning of the expense involved in rescuing a protected areas system on the brink of collapse: the sum represents efforts to conserve about fourteen key protected areas with representation in all biogeographical units (C. Roque, personal communication).

DISCUSSION

Our analyses highlight the skewed regional distribution of large reserves (greater than 1,000 km²), the prevalence of many small reserves per country, and the large number of small reserves that face impending isolation. We thus recommend that the framework for a regional investment strategy contain three key elements: (1) expansion and improved management of existing protected areas of high biological value; (2) establishment of new parks and transfrontier reserves to address gaps in coverage within biogeographic units; and (3) integration of protected areas containing tropical moist forest into regional landscape planning and forest management. Interventions by multi- and bilateral donors and other conservation donors should be evaluated as to how effectively their proposed investments contribute to these three elements. The landscape approach is especially critical for the nearly four hundred small protected areas in the tropical forest zone. We emphasize investments in large conservation units, whatever their legal status or IUCN category, because large reserves have the highest probability of maintaining representative ecosystems, areas of high species richness and endemism, ecological processes, and viable populations of threatened and keystone species (Noss and Cooperrider 1994). Large reserves also minimize effects of local climate change and buffer core areas from the growing resource demands of high-density human populations in rural areas throughout the region.

Results from our analyses identify priorities for investments to conserve tropical forest biodiversity in a network of protected areas at the regional,

Table 10.1 Investments in biodiversity conservation for Indo-Pacific countries by U.S. institutions in 1989, planned investments by the Global Environment Facility (GEF) Operations Program beginning in 1992, and local resources

Category Country	Total area in protection system[a] (km²)	Protected areas containing tropical moist forest (existing and proposed) (km²)	1989 funding U.S. institutions[b,c]	National budget for tropical moist forest conservation ($US)[c]	$US/km²/yr protected habitat: U.S. institutions	$US/km²/yr protected habitat: GEF[d]	$US/km²/yr protected habitat: national budgets	Per capita GNP ($US)
I								
Indonesia		265,983	1,394,244	6,000,000	5.2	9.0	22.6	500
Malaysia		40,914	448,846	5,000,000	11.0	0.0	122.2	2,160
Laos		47,211	—	10,000	—	23.3	0.2	180
Cambodia		25,026	—	—	—	0.0	—	130
Brunei		1,182	—	—	—	0.0	—	17,230
II								
Australia (Qld.)		7,605	—	—	—	0.0	—	14,440
Thailand		56,645	699,449	15,100,000	12.3	35.3	266.6	1,230
Sri Lanka	7,837	6,309	30,430	511,000	4.8	97.0	81.0	430

Table 10.1 (continued)

Category Country	Total area in protection system[a] (km²)	Protected areas containing tropical moist forest (existing and proposed) (km²)	1989 funding U.S. institutions[b,c]	National budget for tropical moist forest conservation ($US)[c]	$US/km²/yr protected habitat: U.S. institutions	$US/km²/yr protected habitat: GEF[d]	$US/km²/yr protected habitat: national budgets	Per capita GNP ($US)
III								
Papua New Guinea		9,164	68,952	925,000	7.5	109.1[f]	100.9	890
Burma		13,040	10,151	1,400,000	0.8	0.0	107.4	<500
Solomon Islands		0	0	—	0.0	—[f]	—	570
Fiji		53	—	—	—	—[f]	—	1,640
IV								
India	134,811	41,550	200,260	1,233,000	4.8	0.0	29.7	340
Vietnam	219,471	6,252	5,500	225,000	0.9	96.0	36.0	220
China		3,539	10,744	645,000	3.0	0.0	—	360
Philippines	2,886	2,395	363,068		151.6	1,670.1	269.3	700

Table 10.1 (continued)

Category Country	Total area in protection system[a] (km²)	Protected areas containing tropical moist forest (existing and proposed) (km²)	1989 funding U.S. institutions[b,c]	National budget for tropical moist forest conservation ($US)[e]	$US/km²/yr protected habitat: U.S. institutions	$US/km²/yr protected habitat: GEF[d]	$US/km²/yr protected habitat: national budgets	Per capita GNP ($US)
Bangladesh		744	15,000	645,000	20.2	0.0	866.9	180
Taiwan		616	—	—	—	0.0	—	7,380
Vanuatu		0	—	—	—	—[f]	—	860

[a] The total area under protection is included only for countries that have large areas under protection containing no tropical moist forest.

[b] From Abramovitz 1991.

[c] For Sri Lanka, India, China, and Taiwan, this value is estimated by multiplying the total budget and the fraction of the area under protection that contains tropical moist forest.

[d] The lifespan of a GEF project was assumed to be five years in order to annualize the figures.

[e] GEF funds are planned for the following countries: Indonesia, $US12 million (not confirmed); Laos, $US5.5 million; Thailand, $US10 million; Sri Lanka, $US3.8 million; Papua New Guinea, $US5 million (not confirmed); Vietnam, $US3 million; Philippines, $US20 million. Data are from UNDP, World Bank, and UNEP 1991. None of the biodiversity projects has been implemented to date.

[f] An undetermined amount will be added. The regional GEF contribution to the South Pacific will be $US8.2 million. Biodiversity projects will be supported in 22 countries, including the 4 covered here.

biogeographic unit, national, and local landscape scales. From each perspective, we suggest courses of action to address outstanding needs of existing areas and gaps in geographical coverage (table 10.2). Although we recommend investments at various scales, these are often hierarchical, so that interventions may have powerful multiplying effects. Priorities are identified by the urgency of threat, the potential for conservation, and the perceived lack of local financial and technical support. At the regional and national scales, priorities identified here should reflect political and socioeconomic constraints and local capacity to use foreign assistance. However, these considerations should serve as secondary filters for refining priorities for biodiversity conservation rather than as the initial template and so are not discussed here.

Regional and National Patterns: Recommendations from the CPTI

At the regional level, the CPTI yields the following priorities, presented by category of threat and potential for investing in biodiversity conservation in tropical moist forest countries:

Category I is the conservation ideal; however, few countries qualify, and those that do, such as Indonesia, often have problems protecting and staffing established reserves. An important target for conservation financing is to ensure that those large, high-priority reserves that are still proposed or are gazetted but essentially "paper" parks, become operational as soon as possible. At present, Laos lacks protected areas and Cambodia has just a few; the CPTI (see fig. 10.4) thus assumes that all proposed areas will be gazetted and under some form of conservation management by the end of the 1990s. Priority investments in these countries should be in rapid assessment surveys to determine the most important areas to protect first. Laos and Cambodia are among the poorest countries in the region and lack technical expertise for management in comparison with other category I countries. Use of trust funds to build infrastructure and train staff would receive high priority. Pressures from neighboring countries to log the forests of Laos and Cambodia require mitigating action and development of a regional forest conservation strategy for Indochina that includes creation of transfrontier reserves (see fig. 10.10).

Countries in category II have placed a relatively large portion of their remaining tropical moist forests under formal protection, but heavy deforestation still occurs outside protected areas. Thailand's forests are rich in species and Sri Lanka's forests are scarce and high in endemics, qualifying these two nations as high-priority countries. Within the 1990s Thailand will likely have the economic means to successfully manage its network of protected areas and will

Table 10.2 Recommendations for an investment portfolio to conserve tropical forests in protected areas of the Indo-Pacific region

Scale	Recommendations	Approach	Priority at this scale
Regional and national	1. For category IV countries: identify remaining large forest tracts for investments, emphasize integration of small reserves in forest matrix where possible, identify isolated remnants of highest biological value	1. Update or design plans for intensive management of small reserves	A
	2. Develop buffer-zone projects, reforestation projects, and plantations to relieve pressure on remaining forests in category II countries	2. Implement community-based approaches, integrated conservation and development projects (ICDPs), surveys	A
	3. Ensure that investments to strengthen management of Indonesian reserves are on a par with Indonesia's role as the largest reservoir of tropical moist forests in the region	3. Trust funds, training programs, surveys	B
	4. Expand coverage of protected areas in category II and III countries to shift them into category I as soon as possible	4. Trust funds, endowments	B
	5. Develop a regional conservation strategy for Indochinese tropical forests, invest in transfrontier reserves, and halt illegal logging	5. Policy development, trust funds, endowments	B
	6. For South Pacific countries with extensive tropical moist forests under customary ownership (e.g., PNG, Solomons), establish large conservation areas promoting innovative clan-based approaches to reserve management	6. ICDPs	C
Biogeographical	1. Increase incorporation of lowland forest into existing and proposed reserves in Sumatra, Irian Jaya, Borneo, and other reservoirs of lowland rainforests	1. Integrate with national conservation strategies, subregional strategies	A
	2. Develop long-term conservation plan for reservoirs of tropical forest (i.e., heavily forested units of category III—see fig. 10.7)	2. Trust funds, regional strategy	B

Table 10.2 (continued)

Scale	Recommendations	Approach	Priority at this scale
Biogeographical (continued)	3. Expand coverage of protected areas to conserve 20% of forest cover in all biounits and emphasize increased protection in biounits with low regional representation	3. Integrate with national conservation strategies, subregional strategies	B
Landscape level	1. Include components that consider expanding management of conservation units beyond reserve boundaries using new approaches to landscape management	1 and 2. GIS analyses, interagency task forces, policy work; support model field projects that highlight a landscape approach and are replicable	A
	2. Incorporate landscape approach into country studies and national conservation strategies		B
Other	1. Support finer-scale CPTIs developed by local conservation planners to set priorities within countries; determine economic potential of protected areas and monitor effectiveness of investments	1. Planning workshops, small grants	A
	2. Increase quality of information available for conservation planning at subnational scale	2. Surveys, biodiversity information network	B
	3. Develop fund for rapid assessment of proposed protected areas and expansions	3. Fund as component of a national conservation	C

Notes: Priority rankings correspond to the numbered recommendations.

A = highest priority, to be undertaken immediately

B = high priority, to be undertaken within 5 years

C = intermediate priority, to be undertaken within 10 years

require assistance largely in the areas of technical training, research, and buffer zone management. Thailand's Huay Kha Khaeng-Thung-Yai Narusan Sanctuary is perhaps the most important large reserve complex in Southeast Asia (see fig. 10.10). Full protection for the isolated forest fragments in Sri Lanka, particularly the Sinharaja forest, is a major priority. Remaining forest tracts that can be integrated into larger conservation matrixes must also be considered a priority. Global Environment Facility support to both countries is timely and addresses critical problems in management of protected areas, although Sri Lanka will require substantially more international support once the capacity to use assistance effectively has increased.

Countries in category III will have relatively large proportions of forest habitat remaining in the year 2000 and thus have the greatest potential for establishing an integrated protected areas system. Category III countries that are high in species richness and endemicity, such as Malaysia, Papua New Guinea, Laos, Burma, and the Solomon Islands, represent important opportunities for establishing national networks of large protected areas. External financing for conservation efforts in Indochina and the South Pacific has been minimal or nonexistent and should receive the highest priority (see table 10.1; Dinerstein, unpublished data). Customary ownership of forested areas in Papua New Guinea and the Solomons will likely preclude the establishment of large government-managed protected areas, but clan-based initiatives and new approaches to conserving large forested tracts should be given highest priority. Excepting Malaysia, countries of category III would greatly benefit by the establishment of conservation trust funds to support the design, survey, and management of protected areas. These countries share similar problems of low absorptive capacity: few trained staff to manage protected areas, the need for extensive surveys of tropical forests in biologically important areas, limited local financing, and lack of incentives to avoid overexploitation of forest resources. Although Malaysia is a newly industrialized country that can afford a strong protected areas system, it shows limited commitment to maintaining extensive tracts of primary forest.

Category IV countries have few protected areas, and little of their forests outside protected areas are expected to survive the next ten years. Most reserves are scattered and insularized, resulting in isolated populations; even if clumped, the reserves may not preserve all habitat diversity (see appendix 10.3 and fig. 10.11 top). Biodiversity considerations indicate that key category IV countries include the Philippines, Vietnam, and China. The Philippines has benefited from major financial assistance for conservation activities, whereas foreign assistance to Vietnam has remained at a trickle.

In sum, action is required to ensure that countries from categories II and III move into category I, and immediate efforts must be taken to halt erosion of biological diversity and establish parks in category IV countries (see table 10.2). Ultimately, investments in biodiversity conservation by the GEF and other donors should be evaluated with this strategy in mind. Regionally, high conservation potential exists in all countries of Indochina (see fig. 10.3)—they contain the large forest blocks needed to upgrade the size of the existing protected areas into the proposed protected areas system, which would contain a more complete representation of tropical moist forest communities. Recent GEF contributions notwithstanding, funding for protected areas in this subregion has been neglected. Indonesia and Malaysia also offer potential for considerable expansion of the protected areas system (see fig. 10.3). Several countries will have to focus on consolidation of existing areas and intensive management to maintain species diversity and viable populations in small reserves (Sri Lanka, Vietnam, Philippines, southern China). Brunei (category I) and Taiwan and Australia (category IV) are the only Indo-Pacific countries in which conservation activities are well financed.

Biogeographic Considerations

Priorities inferred from a biogeographic approach (see figs. 10.5, 10.6b, 10.7a, 10.7b, and 10.9 bottom) provide a check against national analyses, particularly in terms of ensuring the adequate conservation of representative tropical forest ecosystems. From data in figures 10.7a and 10.7b we recommend using the greatest differences between existing and proposed size of protected areas coverage by biounit (that is, length of the dotted line) as a criterion for identifying highest conservation potential. Among the heavily forested units (category III), Sumatra, Irian Jaya, Borneo, and Sulawesi have the highest potential of exceeding or approaching the goal of at least 20% of tropical moist forests in those units under formal protection (see fig. 10.7a). Papua New Guinea may achieve similar protection, but through clan-based forest management schemes not formally considered protected areas by IUCN. Biounits in Indochina (Indochina, South Indochina, Burmese Coast, and the Irrawaddy units) and in Indonesia (Maluku and the Lesser Sundas) require assistance to realize the potential for conservation.

Immediate attention must be given to conservation of biounits that have only a small fraction of the regional total of tropical moist and lowland rainforest and less than 20% of these remaining forest types under conservation management (10 and 8, respectively, for tropical moist forest and lowland rainforest, category IV). High potential for expansion of lowland rainforest conservation occurs in forest-rich units of Sumatra, Irian Jaya, Borneo, the Malay Peninsula, the Burmese coast, and Java and Bali, which contain smaller amounts of the regional total. We recommend that investments in biodiversity conservation be measured against the degree to which they contribute to increased protection of these units. Regionally, the highest priority should be given to conserving the largest tracts of tropical forests, found in the biogeographic subunits of Indonesia and New Guinea.

Landscape-Level Considerations

Isolation and fragmentation of protected areas pose a major long-term threat to the conservation of the tropical forest biota. The CPTI at the landscape level indicates those countries and biounits that contain a large proportion of small isolated protected areas and will require investments to maintain the biota within these small reserves. The CPTI also indicates where investments to expand the conservation effect to surrounding forested areas have the highest potential for success. Landscape management of protected and multiple-use forests provides an important opportunity for extending conservation management far beyond the boundaries of small existing reserves. This can only be accomplished, however, where management procedures in multiple-use forests are truly compatible with conservation and when the concept moves from rhetoric to reality. Landscape management requires the collaborative efforts of several government agencies and a variety of local user groups, often with conflicting interests and goals. Country studies now being prepared with support from UNEP (United Nations Environment Programme) for the Biodiversity and Agenda 21 conventions should identify opportunities where the management of isolated protected areas can be integrated to create a larger regional landscape compatible with forest conservation. Assessments are required to determine where such efforts are working and how they can be replicated around the region.

Additional Funding Needs

Another area that should be targeted for investment is the information gap on biodiversity that exists in many countries. Data on forest cover and biological richness throughout the region are typically available only at scales inappropriate to conduct comparisons across and within regions. The most glaring need is a standardized data set in GIS format, distinguishing primary, secondary, and seriously degraded forests and rates of loss or degradation at the subnational scale. These could then be disaggregated to compare forest cover from a biogeographic perspective. How extensively these data are available by country should be addressed in the country studies being prepared by UNEP.

National biodiversity strategies for each country are essential tools for planning, but few countries have such endeavors under way. Country strategies should include an attempt to set priorities using the variables described in the CPTI. Another important component is a program for rapid assessment of the biological value of reserves proposed for conservation. Strategies should also include new approaches to evaluating the biological value of and threats facing existing reserves so that funds are allocated based on objective criteria rather than political considerations. Finally, the effectiveness of conservation investments requires investigation, particularly the potential for protected areas to be self-supporting through innovative financing mechanisms (such as a levy on sustainably harvested timber in adjacent forests) and the economic potential of ecotourism.

CONCLUSION

Long-term maintenance of the biological integrity of tropical forest ecosystems and communities requires conservation of large and preferably contiguous tracts of forests. Major new investments in biodiversity conservation should be earmarked for expanding the protected areas system of each country and, where possible, expanding the size of biologically important reserves in all countries. Donors must recognize the limited amounts of conservation financing previously available to high-priority countries and reserves identified by our regional analyses and address these deficiencies.

Most tropical moist forest reserves in the region are not large enough to conserve entire ecosystems and maintain minimum viable populations of many larger species. Intensive management is thus required to deal with demographic, genetic, and environmental threats of extinction associated with isolated populations in small reserves. The dilemmas associated with managing numerous small populations will be the legacy conservationists leave for the next generation unless reserves are incorporated into larger conservation units. Without timely foreign assistance, strong local commitment, and improved landscape conservation, the tropical moist forests of the Indo-Pacific will be reduced to a few secure reservoirs and many small remnants unlikely to survive the next century.

Appendix 10.1 Distribution of tropical moist forests and lowland rainforests in the Indo-Pacific region and other pertinent data used to construct the CPTI

Country	Area (km²)	Remaining area of tropical moist forest (km²)	Tropical moist forest in existing and proposed PAs (km²)	Protected areas >1,000 km² containing tropical moist forest existing (N)	proposed (N)	Tropical moist forest in protected areas >1,000 km² (existing and proposed)	Remaining area of tropical moist forest outside protected areas (km²)	Deforestation rate (%)	Remaining TMF outside protected areas in 10 years (km²) (assume existing and proposed)	Remaining area of lowland rainforest (km²)	Lowland rainforest in existing and proposed protected areas (km²)	Lowland rainforest in protected areas >1,000 km² (existing and proposed)	Protected areas with >20% lowland rainforest existing (N)	proposed (N)
Australia	7,618,000	10,516	7,605	2	0	3,588	2,911	0.1	2,882	10,516	2,836	594	12	0
Bangladesh	143,988	9,370	744	0	0	0	8,626	0.9	7,850	5,310	100	0	1	0
Brunei	5,765	4,692	1,182	0	0	0	3,510	1.5	2,984	2,670	1,266	0	6	1
Burma	678,031	311,850	13,040	2	1	6,219	298,810	0.3	289,846	147,340	320	0	1	1
Cambodia	181,940	71,500	25,026	3	3	25,102	46,474	0.3	45,080	55,500	1,500	700	1	0
China	9,597,000	24,200	3,539	1	0	2,000	20,661	0.8	19,421	6,600	0	0	0	0
Fiji	18,235	6,970	53	0	0	0	6,917	0.1	6,848	6,070	60	0	0	0
India	3,166,828	228,330	41,550	3	2	11,044	186,780	0.3	181,177	37,435	2,144	0	12	9
Indonesia	1,919,443	1,179,140	265,983	26	35	224,128	913,157	0.5	867,499	783,170	88,707	76,789	33	35
Laos	236,725	124,600	47,211	0	20	32,112	77,389	1.2	68,102	31,130	–	–	0	0
Malaysia	332,669	200,450	40,914	4	7	19,740	159,536	1.2	140,392	148,280	13,129	9,085	15	8

Appendix 10.1 (continued)

Country	Area (km²)	Remaining area of tropical moist forest (km²)	Tropical moist forest in existing and proposed PAs (km²)	Protected areas >1,000 km² containing tropical moist forest existing (N)	proposed (N)	Tropical moist forest in protected areas >1,000 km² (existing and proposed)	Remaining area of tropical moist forest outside protected areas (km²)	Deforestation rate (%)	Remaining TMF outside protected areas in 10 years (km²) (assume existing and proposed)	Remaining area of lowland rainforest (km²)	Lowland rainforest in existing and proposed protected areas (km²)	Lowland rainforest in protected areas >1,000 km² (existing and proposed)	Protected areas with >20% lowland rainforest existing (N)	proposed (N)
Philippines	300,000	66,020	2,395	0	0	0	63,625	1.0	57,263	13,870	620	0	3	1
Papua New Guinea	462,840	366,750	9,164	2	0	7,742	357,586	0.1	354,010	229,870	641	0	8	0
Solomon Islands	29,790	25,590	0	0	0	0	25,590	0.1	25,334	24,810	0	0	0	0
Sri Lanka	65,610	13,243	6,309	1	0	1,317	6,934	0.6	3,868	1,203	50	0	1	0
Taiwan	35,988	1,660	616	0	0	0	1,044	3.5	981	1,660	0	0	0	0
Thailand	514,000	106,900	56,645	15	5	32,086	50,255	2.7	36,686	54,900	688	0	6	0
Vanuatu	12,189	2,360	0	0	0	0	2,360	0.1	2,336	2,360	0	0	0	0
Vietnam	329,556	56,680	6,252	1	0	1,820	50,428	0.7	46,898	28,040	175	0	2	0
Total		2,810,821	528,228	60	73	366,898	2,282,593		2,159,456	1,590,734	112,236	87,168	101	55

Appendix 10.2 Distribution of tropical moist forest and lowland rainforest among biogeographic units of the Indo-Pacific region

Biogeographic unit	Tropical moist forest (including lowland rainforest)				Protected areas >1,000 km²			Lowland rainforest only				Protected areas with >20% lowland rainforest cover		
	Remaining area (km²)	Existing protected areas (km²)	Proposed protected areas (km²)	Existing and proposed protected areas (km²)	Existing (N)	Proposed (N)	Unknown (N)	Remaining area	Existing protected areas (km²)	Proposed protected areas (km²)	Existing and proposed protected areas (km²)	Existing (N)	Proposed (N)	Unknown (N)
Western Ghats	32,260	2,139	585	2,724	1	0	0	14,000	1,196	185	1,381	8	0	0
Ceylon wet zone	50	0	50	50	0	0	0	50	0	50	50	1	1	0
Bengal/Assam	36,800	3,637	260	3,897	1	0	0	22,000	920	100	1,020	3	1	0
Eastern Ghats	2,600	852	0	852	0	0	0	0	0	0	0	0	0	0
Burmese coast	22,368	210	3,880	4,090	0	0	0	8,029	210	3,545	3,755	1	0	2
South Indochina and Annam	72,391	8,383	3,565	11,948	2	4	0	15,307	1,988	375	2,363	9	0	0
South China	19,899	177	196	373	0	0	0	0	0	0	0	0	0	0
Irrawaddy	28,601	0	3,217	3,217	2	1	1	0	0	0	0	0	0	0
Indochina	87,657	14,203	6,030	20,233	15	23	0	4,446	13	0	13	0	0	0
Andaman Island	3,100	270	1,140	1,410	0	0	0	2,100	108	378	486	2	9	0
Taiwan	7,537	970	0	970	0	0	0	0	0	0	0	5	0	2
Malay Peninsula	82,897	22,607	5,521	28,128	4	4	0	42,982	7,039	3,098	10,137	5	5	2
Sumatra	167,940	20,797	15,438	36,235	10	3	0	72,129	8,075	10,388	18,463	10	12	10
Java and Bali	11,418	2,430	4,973	7,403	0	0	0	1,398	603	496	1,099	4	5	2
Borneo and Palawan	508,038	23,278	64,385	87,663	10	18	3	355,620	15,193	47,552	62,745	26	17	9

Appendix 10.2 (continued)

Biogeographic unit	Tropical moist forest (including lowland rainforest)				Protected areas >1,000 km²			Lowland rainforest only				Protected areas with >20% lowland rainforest cover		
	Remaining area (km²)	Existing protected areas (km²)	Proposed protected areas (km²)	Existing and proposed protected areas (km²)	Existing (N)	Proposed (N)	Unknown (N)	Remaining area	Existing protected areas (km²)	Proposed protected areas (km²)	Existing and proposed protected areas (km²)	Existing (N)	Proposed (N)	Unknown (N)
Lesser Sunda Islands	19,194	1,565	4,370	5,935	0	0	0	0	0	0	0	0	0	0
Sulawesi	106,653	6,008	11,090	17,098	3	4	0	20,032	1,313	1,450	2,763	5	5	0
Philippines	41,691	1,725	880	2,605	0	0	0	14,637	550	0	550	3	0	0
Maluku	56,070	2,095	21,605	23,700	1	5	0	44,160	1,800	10,200	12,000	1	0	0
Irian Jaya	354,360	42,925	35,907	78,832	5	9	0	232,610	42,925	35,907	78,832	3	1	0
Papua New Guinea	366,750	9,164	705	9,869	2	0	0	229,870	9,164	0	9,164	8	0	0
Solomon Islands	25,590	0	0	0	0	0	0	24,810	0	0	0	0	0	0
Fiji	6,610	53	0	53	0	0	0	6,070	66	0	66	0	0	0
Vanuatu	2,360	0	0	0	0	0	0		0	0	0	0	0	0
Queensland	10,516	7,605	0	7,605	2	0	0	10,000	2,836	0	2,836	12	0	0
Total	2,073,350	171,093	183,797	354,890	58	71	4	1,120,249	93,999	113,724	207,723	101	55	25

Appendix 10.3 Spatial distribution of protected areas containing tropical moist forest in the Indo–Pacific region

Country	Protected areas			Existing PAs with >1 nearest neighbor		Median distance to nearest neighbor (km)		PAs in TMF matrix	
	Total (N)	Existing (N)	Proposed (N)	Existing (N)	Existing and proposed (N)	Existing	Existing and proposed	Existing (N)	Proposed (N)
Australia	14	14	0	3	3	75	75	8	0
Bangladesh	5	5	0	2	2	45	45	4	0
Brunei	7	6	1	4	4	26	26	5	1
Burma	13	7	6	0	0	200	140	6	4
Cambodia	8	5	3	1	1	80	50	5	3
China	8	8	0	0	0	75	75	1	0
India	130	52	78	25	25	25	25	39	52
Northeast India	58	14	44	5	5	65	30	11	31
Western Ghats	57	34	23	19	19	25	25	25	17
Indonesia	215	99	116	14	14	40	30	55	76
Irian Jaya	31	13	18	0	0	30	25	8	16
Java and Bali	28	12	16	0	0	45	45	0	0
Kalimantan	37	14	23	0	5	60	43	13	22
Lesser Sundas	16	11	5	0	0	75	68	0	0
Maluku	19	3	16	0	0	island	50	1	9
Sulawesi	32	17	15	1	1	43	25	10	12
Sumatra	52	29	23	13	18	25	25	23	17
Laos	49	0	49	0	10	—a	30	0	29
Northern Laos	18	0	18	0	0	—a	38	0	6
Southern Laos	31	0	31	0	10	—a	25	0	23

Appendix 10.3 (continued)

Country	Protected areas			Existing PAs with >1 nearest neighbor		Median distance to nearest neighbor (km)		PAs in TMF matrix	
	Total (N)	Existing (N)	Proposed (N)	Existing (N)	Existing and proposed (N)	Existing	Existing and proposed	Existing (N)	Proposed (N)
Malaysia	37	18	19	2	4	60	40	13	19
Malay Peninsula	15	6	9	0	0	60	60	4	9
Sabah	7	7	0	0	0	35	35	4	0
Sarawak	15	5	10	2	4	40	20	5	10
Philippines	11	9	2	0	0	155	155	0	0
Central Island	3	2	1	0	0	island	island	0	0
Luzon	4	4	0	0	0	155	105	0	0
Mindanao	2	2	0	0	0	210	210	0	0
Palawan	2	1	1	0	0	190	190	0	0
Papua New Guinea	12	12	0	0	0	75	75	5	0
Sri Lanka	19	19	0	8	8	20	20	6	0
Taiwan	4	3	1	0	0	45	45	1	0
Thailand	91	69	22	21	23	25	25	35	12
Vietnam	27	27	0	3	3	40	30	16	0

Note: The protected areas used in this analysis are those for which maps were available (Collins et al. 1991).

[a] For Laos, the median distance to the nearest neighbor was calculated from proposed protected areas.

REFERENCES

Abramovitz, J. N. 1991. *Investing in Biological Diversity*. World Resources Institute, Washington, D.C.

Ashton, P. 1969. Speciation among tropical forest trees: Some deductions in the light of recent evidence. *Biol. J. Linn. Soc.* (London) 1: 155–196.

Collins, N. M., J. A. Sayers, and T. C. Whitmore. 1991. *The Conservation Atlas of Tropical Forests: Asia and the Pacific*. Macmillan, London.

Dinerstein, E., and E. D. Wikramanayake. 1993. Beyond hotspots: How to prioritize investments to conserve biodiversity in the Indo-Pacific region. *Cons. Biol.* 7: 53–65.

IUCN (International Union for the Conservation of Nature and Natural Resources). 1986a. *Review of the Protected Areas System in the Indo-Malayan Realm*. IUCN, Gland, Switzerland.

————. 1986b. *Review of the Protected Areas System in Oceania*. IUCN, Gland, Switzerland.

————. 1990. *United Nations List of National Parks and Protected Areas*. IUCN, Gland, Switzerland.

————. 1991. *IUCN Directory of Protected Areas in Oceania*. IUCN, Gland, Switzerland.

MacKinnon, K. 1990. *Biodiversity Action Plan for Indonesia*. Bappenas, Ministry of Population and Environment, and World Bank, Bogor Agricultural University, Bogor.

McNeely, J. A., K. R. Miller, W. V. Reid, R. A. Mittermeier, and T. B. Werner. 1990. *Conserving the World's Biological Diversity*. IUCN/WRI/CI/WWF–US/World Bank, Washington, D.C.

Noss, R. F., and A. Y. Cooperrider. 1994. *Saving Nature's Legacy*. Island Press, Washington, D.C.

Redford, K. H., and J. G. Robinson. 1991. Park size and the conservation of forest mammals in Latin America. In *Latin American Mammalogy*. M. A. Mares and D. J. Schmidley, editors. University of Oklahoma Press, Norman.

UNDP (United Nations Development Programme), World Bank, and UNEP (United Nations Environmental Programme). 1991. Report by the chairman to the participants' meeting to discuss the Global Environment Facility. Volume 2. World Bank, Washington, D.C.

Whitmore, T. C. 1984. *Tropical Rain Forests of the Far East*. Second edition. Oxford University Press, Oxford.

Wilson, E. O. 1988. *Biodiversity*. National Academy Press, Washington, D.C.

World Bank. 1990. Mapping of the natural conditions of the Philippines. In *Philippines Forestry, Fisheries, and Agricultural Resource Management Study, 1989*. World Bank document no. 7388–PH, Solna, Sweden.

World Conservation Monitoring Centre. 1992. *Global Biodiversity: Status of the Earth's Living Resources*. Chapman & Hall, London.

World Resources Institute. 1990. *World Resources—1990–91: A Guide to the Global Environment*. Oxford University Press, New York.

11

IDENTIFYING SITES OF GLOBAL IMPORTANCE FOR
CONSERVATION: THE IUCN/WWF CENTRES OF PLANT
DIVERSITY PROJECT
Stephen D. Davis

Plants are the basis of most terrestrial ecosystems; most animals, including humans, depend on them as sources of food. Worldwide, tens of thousands of species of higher plants, and several hundred species of lower plants, provide a host of other products for human use, including fuel, fibers, oils, medicines, dyes, tannins, and forage crops. In the tropics alone 25,000–30,000 species are estimated to be in use (Heywood 1992, 1993), and up to 25,000 species have been employed in traditional medicines. Collectively, plants also provide many valuable ecological services, such as protecting watersheds, stabilizing slopes, improving soils, moderating climate, cycling nutrients, and providing habitat for animals.

Plant life throughout the world, and especially in the tropics, is under serious threat as habitats are destroyed or modified. Approximately 40% of the land that can support closed tropical forest now lacks forest cover, primarily because of human actions of one form or another (Wilson 1988). Many parts of the tropics are likely to lose their forests within the next half century (Dinerstein, Wikramanayake, and Forney, this volume; Myers 1989; Raven 1988). The loss of these forests will inevitably result in species extinctions on a large scale. Just how many species are at risk is a matter of some debate within scientific circles (see, for example, Whitmore and Sayer 1992) and depends on predictions of the extent of habitat destruction, particularly of primary forests, and how many individual species may be able to survive in fragmented, modified, or secondary habitats. Using the theory of island biogeography, which predicts that a 90% reduction in habitat size will lead to the extinction, or near extinction, of half of its species, Raven (1987) has estimated that as many as 60,000 vascular plant species (about 25% of the world's vascular plant flora) could either become extinct or have their populations seriously reduced by 2050 if present trends continue. Approximately 17,000 vascular plants (7% of the earth's vascular plant species) could become extinct in just ten critical areas, or "hotspots," covering 0.2% of the earth's land surface, according to analyses of tropical floras by Norman Myers (1988). Whether or not one accepts these specific predictions, there can be little doubt that the greatest potential loss of plant diversity is in the tropics.

It is against this background that the International Union for Conservation of Nature and Natural Resources (IUCN—the World Conservation Union) and the World Wide Fund for Nature (WWF) began a major international collaborative project in 1989, the Centres of Plant Diversity (CPD) project, with funding by the European Commission and the Overseas Development Administration of the United Kingdom. The CPD project aims to identify sites of global significance for plant conservation.

IDENTIFYING AREAS OF HIGH DIVERSITY AND ENDEMISM

Many studies intended to identify priorities for conservation have focused on the large numbers of species endemic to certain areas, most often with an emphasis on animals, particularly large vertebrates. A recent study by Bird Life International (formerly known as the International Council for Bird Preservation, or ICBP) uses restricted-range bird species (defined as those species with known breeding ranges of less than 50,000 km^2) to identify 221 Endemic Bird Areas (EBAs) (ICBP 1992; Stattersfield et al., in prep.). Birds were selected for two principal reasons: they have dispersed to, and diversified in, all regions of the world, and they occur in virtually all habitat types and altitudinal zones. These arguments apply equally well to plants, which also collectively provide the habitat for the vast majority of these bird and other terrestrial species.

Attempts have been made to identify countries of "megadiversity" (Mittermeier and Werner 1988; Raven 1987) and hotspots (Myers 1988, 1990) of floristic diversity. Raven (1987), for example, has highlighted such areas as Madagascar, lowland western Ecuador, and the Atlantic forests of Brazil as most in need of conservation attention, and Myers (1988, 1990) has identified a number of hotspots, areas that have exceptional concentrations of species and high levels of endemism and are under severe threat of destruction. Most of these hotspots are regions that contain tropical forests (table 11.1).

A number of authors would like to get beyond hotspot analyses to assess priorities within regions. Several approaches have been attempted. Vane-Wright, Humphries, and Williams (1991), for example, have introduced an index based on the taxonomic uniqueness of species and genera within a region;

Table 11.1 Hotspot areas

Region	Areas with tropical forests	Areas without tropical forests
Africa	Eastern Arc Forests of Tanzania Madagascar Southwestern Ivory Coast	Cape Floristic Province of South Africa
Australia	Queensland	Southwestern Australia
Indian subcontinent	Southwestern Sri Lanka Western Ghats of India	Eastern Himalaya
North America		California Floristic Province
Pacific Ocean Islands	Hawaii New Caledonia	
South America	Atlantic Coast of Brazil Colombian Chocó Uplands of Western Amazonia Western Ecuador	Central Chile
Southeast Asia (Malesia)	Northern Borneo Peninsular Malaysia Philippines	

Sources: Myers 1988, 1990.

they advocate the use of this index in determining priorities for conservation. Dinerstein and Wikramanayake (1993) present another approach. They and Forney (this volume) have developed a Conservation Potential/Threat Index (CPTI), which forecasts how deforestation in the 1990s will affect the conservation and establishment of forest reserves by comparing biological richness with reserve size, size of protected area, remaining forest cover, and deforestation rates in twenty-three Indo-Pacific countries.

These approaches are welcome, but no matter how sophisticated the methodology used for selecting priorities for conservation, the success of any conservation program ultimately depends on the practicalities of implementation. Implementation, in turn, requires consideration of a region's political and social dimensions. One principle adopted in the CPD project has been to involve local and regional experts and, wherever possible, national governmental and nongovernmental conservation bodies in identifying key sites for plant conservation.

The idea of preparing a world survey of the centers of plant diversity grew out of the work of the Threatened Plants Unit (TPU) of the IUCN Conservation Monitoring Centre (now the World Conservation Monitoring Centre) in the 1980s. The TPU was concerned both about the rapid rate of loss of habitats worldwide and about the difficulty of applying IUCN Red Data Book categories to many plant species in the tropics. These difficulties arose primarily as a result of the number of species in tropical floras and the poor state of botanical knowledge of many tropical areas (Campbell and Hammond 1989), and because many plant species in tropical forests have scattered distributions. Attempts to assign Red Data Book categories to individual species in order to assess priorities for conservation were ineffective in the tropics; an alternative strategy was needed.

The approach finally agreed on, after much discussion and debate within the Joint IUCN-WWF Plant Advisory Group (which helped to guide the direction of TPU research), was to develop a list of all major sites and vegetation types considered to be of international importance for conservation based on species richness and endemism. The project was named Centres of Plant Diversity (CPD). The results of the project will be published as three volumes.

The objectives of CPD are: to identify which areas around the world, if conserved, would safeguard the greatest number of plant species; to document the many benefits, both economic and scientific, that conservation of those areas would bring to society and to outline the potential value of each for sustainable development; and to outline a strategy for the conservation of the areas selected.

THE CRITERIA AND METHODOLOGY USED FOR SELECTING SITES

In order to be selected, sites and vegetation types had to have one or both of the following two characteristics: the area had to be evidently species-rich, even though the number of species present might not be accurately known; and the area had to be known to contain a large number of endemic species. Other characteristics that were considered in site selection were the presence of any or all of the following: an important gene pool of plants of actual or potential value to humans, a diverse range of habitat types, a significant proportion of species adapted to special edaphic conditions, or an imminent threat of large-scale devastation.

The concept of the CPD project is related to that of the work by crop geneticists in selecting centers of origin and diversity of crop plants—the so-called Vavilov Centers of Crop Genetic Diversity (Hawkes 1983). The main criteria for selecting CPD sites, however, are high plant species diversity and endemism, with habitat diversity and the presence of important gene pools of plants as secondary considerations. The CPD project has also documented the benefits of conserving the areas and outlined the potential of each for sustainable development in line with the principles of the World Conservation Strategy (IUCN 1980) and the Convention on Biological Diversity.

More than four hundred botanists, ecologists, and conservationists from more than one hundred institutions have advised and collaborated in the project, and many workshops have been held to review lists of sites. The result is a list of 234 global priority sites treated in data sheets in the CPD volumes. The data sheets contain information on the geography, flora, vegetation, useful plants, social and environmental values, threats, and conservation of each selected site. In a few cases, a partial and preliminary economic assessment of the value of plant resources is included. This assessment may be based on the number of visitors to an area or the amount of wild plant resources gathered from a site. In addition to the top priority sites treated in data sheets, nearly six times this number of sites are listed with summary data.

Most mainland CPD sites have (or are believed to have) in excess of 1,000 vascular plant species, of which at least 100 (10%) are endemic either to the site (strictly endemic) or to the phytogeographical region. In many cases, the number of regional endemics is much higher than 10% of the flora, and all sites have at least some strictly endemic species. Often a significant proportion of the total flora is endemic to the site.

Islands have been selected using somewhat different criteria from those used for mainland sites. Many islands have depauperate floras compared with continental areas, but they often have a very high level of endemism. Restricting the selection to floras of 1,000 species or more would have led to the omission of some important areas rich in endemics. The principal selection criteria for islands were therefore adjusted: to qualify for selection an island flora must contain at least 50 endemic species or at least 10% of the flora must be endemic to the island or island group. In fact, several sites were selected on many of the larger islands in Southeast Asia (Malesia) to cover a range of vegetation and community types.

Although a degree of subjectivity was inevitable in deciding which of a number of similar sites would be included in the CPD project, in some parts of Southeast Asia, the choice was relatively easy. Mt. Kinabalu, the focal point of Kinabalu Park in Sabah, Malaysia, and the highest mountain between the

Himalayas and New Guinea, has an exceedingly rich flora of more than 4,000 vascular plant species, and possibly more than 4,500 species, in an area of 753.7 km². Kinabalu has one of the richest concentrations of endemic species in the world and is thus an obvious choice for selection as a top priority (data sheet) site. Java presents another clear case. Here, almost all the lowland forest has been cleared, so the choice of sites is mostly restricted to upland volcanic areas, such as the twin peaks of Gede-Pangrango National Park, which support mostly montane and submontane rain forest, subalpine forest, and grass plains.

In many parts of the region, the choice of sites was not as easy. Certain substrates, such as limestone and ultramafic rocks, often support an endemic-rich flora (Proctor, this volume), but individual outcrops are often quite small and scattered over a wide area. For example, the three hundred limestone hills in Peninsular Malaysia collectively cover an area of some 260 km² and support a flora of more than 1,300 vascular plant species. About 10% of these species are endemic to Peninsular Malaysia and entirely restricted to limestone. As a result, protecting one site, or a few sites, will not adequately conserve the flora. The approach adopted in CPD is to treat such specialized floras as one unit and to recommend a network of sites to protect the flora.

The most difficult problems in site selection are in areas like Papua New Guinea where the flora is exceedingly rich and poorly known. Papua New Guinea has a flora of approximately 15,000–20,000 vascular plant species (R. J. Johns 1992, personal communication), of which 70–80% are endemic. Any sizable tract of native vegetation is therefore likely to contain endemics, making the selection of priorities extremely difficult. In practice, site selection for CPD has relied heavily on the field knowledge of only a few botanists and therefore must be considered preliminary. It is hoped that further identification and documentation of important plant sites will be forthcoming in all areas of the world.

ANALYSIS OF SITE INFORMATION

The locations of the 234 sites selected as top priorities are shown in fig. 11.1. The number of data sheet sites selected for each major region of the world corresponds broadly to the richness of the flora in that region, although it must be remembered that individual sites vary enormously in size. Seventy-five sites have been selected as priorities for Southeast Asia, China and East Asia, and the Indian subcontinent, of which 66 contain tropical forests. Of these, 41 occur in Southeast Asia (fig. 11.2 and appendix 11.1).

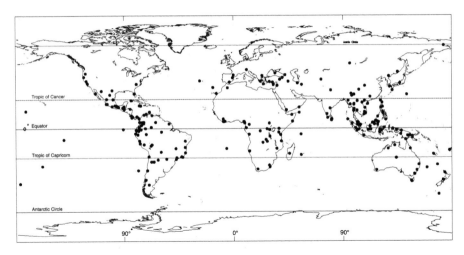

FIGURE 11.1   The locations of the 230 CPD sites selected as top priorities.

Conservation Status

Of the 66 tropical Asian sites selected for inclusion in the CPD project, 27 (41%) enjoy full legal protection, an encouraging figure (table 11.2). However, 8 sites (12%) have no legal protection, and in a further 31 sites (47%) much less than 50% of their area is included within a network of protected areas. This indicates that the present coverage of legally protected areas is inadequate to capture the full range of vegetation types and areas of greatest floristic richness within the regions concerned. To put these figures into context, worldwide fewer than one in four of the data sheet sites (21%) are legally protected in full, and only about one-third of the selected sites (35%) have more than 50% of their areas occurring within existing protected areas. For tropical Asia, the respective percentage figures are slightly higher (41% and 44%), but a considerable proportion of those sites in tropical Asia that are officially protected are not effectively managed, being subject to degradation ranging from logging and slash-and-burn agriculture to clearance for tourist developments. Thus, only 6 CPD sites throughout tropical Asia are categorized under the broad heading of "safe or reasonably safe" (table 11.3).

Of 41 CPD sites in Southeast Asia, just 3 (7%) are considered to be safe or reasonably secure. The protected area systems need to be extended and strengthened if the strategy of protecting biodiversity through setting aside and conserving areas of high floristic and ecological richness is to succeed.

Almost all the tropical sites that are protected (at least on paper) suffer from a lack of adequate funding and trained staff. These shortages affect the

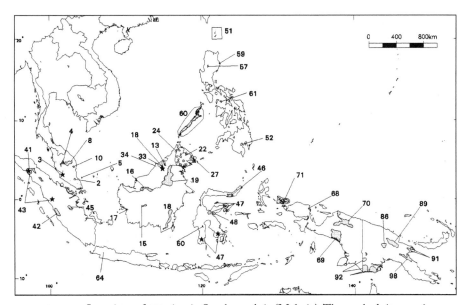

FIGURE 11.2   Locations of CPD sites in Southeast Asia (Malesia). The marked sites receive
data sheet treatment in the CPD project and are considered global priority sites for plant conservation. The site numbers correspond to the site codes used in the CPD project. Many more sites
(not indicated on the map) are identified in the CPD project as regionally important. Stars and
triangles indicate scattered sites treated as one unit. **Malaysia (Peninsular Malaysia):** *2* =
Endau-Rompin State Parks; *3* = limestone flora of Peninsular Malaysia; *4* = montane flora of
Peninsular Malaysia; *5* = Pulau Tioman; *8* = Taman Negara; *10* = Trengganu Hills. **Brunei
Darussalam:** *13* = Batu Apoi Forest Reserve, Ulu Temburong. **Indonesia (Kalimantan):** *15* =
Bukit Raya and Bukit Baka; *17* = Gunung Palung; *19* = Sungai Kayan-Sungai Mentarang Nature
Reserve. **Indonesia and Malaysia:** *16* = Lanjak Entimau, Batang Ai, Gunung Bentuang dan
Karimun; *18* = limestone flora of Borneo. **Malaysia (Sabah):** *22* = East Sabah lowland and hill
dipterocarp forests; *24* = Kinabalu Park; *27* = northeast Borneo ultramafic flora. **Malaysia and
Brunei:** *33* = Gunung Mulu National Park, Medalam Protected Forest, Labi Hills, Bukit Teraja,
Ulu Ingei, Sungei Ingei. **Malaysia (Sarawak):** *34* = Lambir Hills. **Indonesia (Sumatra):** *41* =
Gunung Leuser; *42* = Kerinci-Seblat National Park; *43* = limestone flora of Sumatra; *45* =
Tigapuluh Mountains. **Indonesia (Sulawesi):** *46* = Dumoga-Bone National Park; *47* = lime-
stone flora of Sulawesi; *48* = ultramafic flora of Sulawesi; *50* = Pegunungan Latimojong.
Philippines: *51* = Batan Islands; *52* = Mt. Apo; *57* = Mt. Pulog; *59* = Palanan Wilderness Area; *60*
= Palawan; *61* = Sibuyan Island. **Indonesia (Java):** *64* = Gede-Pangrango National Park.
Indonesia (Irian Jaya): *68* = Arfak Mountains; *69* = Gunung Lorentz; *70* = Mamberamo-
Pegunungan Jayawijaya; *71* = Waigeo. **Papua New Guinea:** *86* = Mt. Giluwe, Tari Gap, Doma
Peaks; *89* = Bismarck Falls, Mt. Wilhelm, Mt. Otto, Schrader Range, Mt. Hellwig, Gahavisuka;
91 = Huon Peninsula (Mt. Bangeta, Rawlinson Range, Cromwell Ranges, Sialum Terraces); *92* =
Southern Fly Platform; *98* = Menyamya, Aseki, Amungwiwa, Bowutu Mts., Lasanga Island.

security of the areas. Boundaries may not be properly marked or policed, and
valuable resources, such as rattans and timber trees, may be illegally exploited in
an uncontrolled manner that is perhaps unsustainable in the long term. Some
areas lack effective management plans to protect or use important genetic re-

Table 11.2   The conservation status of top-priority tropical Asian CPD sites

Region	CPD sites (N)				Sites (N)
	0% protection	*>0–50% protection*	*> 50–100% protection*	*100% protection*	
China and East Asia	0	6	0	7	13
Indian subcontinent	2	8	0	2	12
Southeast Asia	6	17	2	16	41
Total	**8**	**31**	**2**	**25**	**66**

Note: The designation of forest reserve does not necessarily confer any degree of protection on a site's flora. If information gathered during the CPD project indicates that no protection is afforded, especially if large-scale logging is imminent, the percentage of the site that is protected has been adjusted accordingly.

Table 11.3   Degree of threat to tropical Asian CPD sites

Region	Safe or reasonably safe (N)	Partly safe but some areas threatened (N)	Vulnerable or at risk (N)	Threatened (N)	Severely threatened (N)
China and East Asia	2	1	1	5	4
Indian subcontinent	1	6	0	3	2
Southeast Asia	3	7	12	15	4
Total	**6**	**14**	**13**	**23**	**10**

Note: For details on individual sites, see appendix 11.1.

sources. More outreach programs need to be developed to involve local people in managing and protecting the forests and to ensure sustainable use of plant resources. In addition, land zoning within and around the areas is needed to ensure that conservation of biodiversity is balanced against other demands on natural resources. Effective conservation of the CPD sites will depend, therefore, on adequate funding and the political will to establish more protected areas where this is necessary and to ensure that all protected areas are effectively managed.

CONCLUSION

The CPD project has moved beyond "hotspot" analyses to identify sites of global botanical importance. More work is needed by national programs to identify sites that are locally important. The opportunities for conservation are rapidly disappearing. Governments, aid agencies, and conservation organizations must act to achieve the conservation of the areas selected as CPD sites now, before it is too late.

Appendix 11.1 Tropical Asian centers of plant diversity: Summary information on sites selected for data sheet treatment

China and East Asia

CHINA

Site	Type[a]	Area[b] (km²)	Altitude (m)	Flora[c] (N)	Examples of useful plants	Vegetation	Protected areas[d]	Threats[e]	Assessment[f]
Tropical forests of Hainan Island	V	33,920	0–1,867	4,200–4,500	>2,900 species used locally; timber trees, medicinal plants, rattans, wild litchi	Tropical lowland seasonal rainforests and monsoon forests, montane seasonal rainforests, mangroves, savanna	51 NRs cover c. 1,500 km² (including marine reserves)	Clearance for cultivation, illegal logging, over-collection of medicinal plants	Severely threatened; reserve coverage adequate, but lack of funds and trained personnel
Limestone region, Zhuang Autonomous Region	F	20,000	100–1,300	2,500–3,000	>1,000 species used locally; timbers, medicinal plants, bamboos, rattans, ornamentals	Seasonal rainforests, montane evergreen, broadleaved, and limestone forests	11 NRs (4,000 km²)	Illegal timber cutting, lack of conservation awareness by local people	Protected areas need effective management; severely threatened
Xishuangbanna region, Yunnan	F	19,690	500–2,429	4,000–4,500	>800 medicinal plants, 128 timber trees, bamboos, rattans, fruits	Tropical lowland and montane seasonal rainforests, evergreen dipterocarp forests, monsoon forests, montane evergreen and broadleaved forests	5 NRs (2,416 km²)	Slash-and-burn agriculture, clearance for plantation crops, illegal logging, colonization	Severely threatened; reserves cover major plant-rich sites, but threats continue and management is inadequate

Appendix 11.1 (continued)

Site	Type[a]	Area[b] (km²)	Altitude (m)	Flora[c] (N)	Examples of useful plants	Vegetation	Protected areas[d]	Threats[e]	Assessment[f]
TAIWAN									
Kenting NP	S	177 (land)	0–526	1,350	Timber trees, medicinal plants, legumes, ornamentals	Evergreen broadleaved rainforests, semideciduous and littoral forests, grassland, scrub	NP	Tourism, grazing, plant collecting, military facilities	Generally well protected, some parts under threat
THAILAND									
Doi Chiang Dao WS	S	521	<800–2,175	1,200	Teak (logged out), edible fruits	Lowland mixed evergreen–deciduous and dipterocarp-oak forests, montane forests	WS	Fire	Vulnerable
Doi Suthep-pui NP	S	261	360–1,685	2,063[g] taxa	Timber trees, bamboos, fruit trees, medicinal and ornamental plants, mushrooms	Monsoon forests, including lowland dipterocarp-oak and mixed evergreen–deciduous forests, pine forests	NP, BR (whole area)	Encroachment, tourism, road building, tree cutting, fire, overcollection of ornamentals	Lack of resources for full implementation of conservation measures; threatened
Khao Yai NP	S	2,168	400–1,351	2,000–2,500	Medicinal plants, rattans	Moist evergreen forests, dry evergreen and mixed deciduous forests, grassland	NP	Encroachment of agriculture, illegal logging, overcollection of forest products, dam construction	Threatened around perimeter; conservation and development program under way

Appendix 11.1 (continued)

Site	Type[a]	Area[b] (km²)	Altitude (m)	Flora[c] (N)	Examples of useful plants	Vegetation	Protected areas[d]	Threats[e]	Assessment[f]
Thung Yai-Huai Kha Khaeng WHS	S	12,000	200–1,811	>2,500	Crop relatives, ornamentals	Lowland evergreen and deciduous forests, evergreen montane forests	WS, WHS	Encroachment of agriculture, logging, fire, potential dam construction	Management underresourced; threatened
VIETNAM									
Bach Ma-Hai Van	S	600	0–1,450	2,500	200 timber trees, 108 medicinal plants, ornamentals, fibers, rattans, edible fruits	Lowland evergreen forests, tropical montane evergreen forests	NP (220 km²), 2 environment protection forests	High local population density, felling of timber trees, fuelwood cutting	Threatened
Cat Tien BR	S	1,372	60–754	2,500	200 timber trees, 120 medicinal plants, rattans, bamboos, orchids	Lowland evergreen and semi-deciduous forests, freshwater swamps, bamboos	BR includes NP (379 km²), rhino sanctuary (360 km²)	Illegal logging, over-exploitation of rattans and resins	Reasonably secure
Cuc Phuong NP	S	300	200–636	1,980[g]	Timber trees, medicinal plants, bamboos, ornamentals	Lowland evergreen forests, including limestone forests, semi-deciduous forests	NP (222 km²)	Illegal felling of timber trees, fuelwood cutting	Reasonably secure
Langbian-Dalat Highland	S	4,000	1,400–2,289	2,000	100 medicinal plants, ornamentals (especially orchids), resin trees, rattans	Pine forests, tropical montane evergreen forests, subtropical montane forests	3 NRs (535 km²)	Logging, slash-and-burn cultivation, fire	Severely threatened

Appendix 11.1 (continued)

Site	Type[a]	Area[b] (km²)	Altitude (m)	Flora[c] (N)	Examples of useful plants	Vegetation	Protected areas[d]	Threats[e]	Assessment[f]
Yok Don NP	S	650	200–482	1,500	150 timber trees, tannins, resin trees, edible fruits, ornamentals	Dry dipterocarp forests, lowland semi-evergreen forests, riverine forests	NP (580 km²)	Illegal logging, hunting, forest fires	Threatened
Indian Subcontinent INDIA									
Agastyamalai Hills	F	2,000	67–1,868	2,000	Medicinal herbs, timber trees, bamboos, rattans, crop relatives	Tropical dry and wet forests	3 sanctuaries and a number of RFs protect c. 919 km²; proposed BR	Clearance for hydroelectric projects, plantations, tourism, fire, grazing	Some areas threatened
Andaman and Nicobar Islands	F	8,249	0–726	2,270	Rice and pepper relatives, timber trees, rattans	Tropical evergreen, semi-evergreen and moist deciduous forests, beach forests, bamboo scrub, mangroves	c. 500 km² land protected in 6 parks, 94 sanctuaries (mostly small islets); part of Great Nicobar proposed as BR, proposed North Andaman BR	Population pressure, logging, hydroelectric schemes, clearance for agriculture	Some areas severely threatened; some areas reasonably safe at present; protected area coverage inadequate
Nallamalai Hills	F	7,640	300–939	758[g]	Medicinal plants, rice and pepper relatives, bamboos, timbers	Dry & moist deciduous forests, dry evergreen forests, scrub	WS covers 357 km²	Forest clearance, bamboo cutting for paper, fire, fuelwood cutting	Severely threatened

Appendix 11.1 (continued)

Site	Type[a]	Area[b] (km²)	Altitude (m)	Flora[c] (*N*)	Examples of useful plants	Vegetation	Protected areas[d]	Threats[e]	Assessment[f]
Namdapha	S	7,000	200–4,578	5,000	Banana, citrus, and pepper relatives, timber trees, ornamentals	Tropical evergreen and semi-evergreen rainforests, alpine vegetation	NP (1,985 km²); whole area proposed as BR with core of c. 2,500 km²	Shifting cultivation, refugees and other illegal settlers, illegal timber felling	Core area not threatened; perimeter areas under threat
Nilgiri Hills	F	5,520	250–2,000	3,240	Medicinal plants, timber trees, fruit-tree relatives	Tropical evergreen rainforests, tropical dry thorn forests, montane shola forests, and grassland	Several NPs (including Silent Valley NP—89.5 km²), sanctuaries, and RFs; proposed BR	Forest clearance for timber, plantations, roads, development projects	Some areas threatened
MYANMĀ (BURMA) Bago (Pegu) Yomas	F	40,000	100–821	2,000	Timber trees (especially teak)	Wet evergreen dipterocarp forests, moist and dry teak forests, bamboo scrub	Proposed NP (146 km²), some RFs	Slash-and-burn cultivation, road building, dam construction	Protection needs strengthening; threatened
Natma Taung (Mt. Victoria) and Rongklang Range (Chin Hills)	F	25,000	500–3,053	2,500	Gingers, peppers, ornamentals (e.g., orchids)	Tropical and temperate semi-evergreen forests, subtropical evergreen forests, savanna, alpine vegetation	Proposed NP (364 km²)	Slash-and-burn agriculture	Threatened

Appendix 11.1 (continued)

Site	Type[a]	Area[b] (km²)	Altitude (m)	Flora[c] (N)	Examples of useful plants	Vegetation	Protected areas[d]	Threats[e]	Assessment[f]
North Myanmā	F	115,712	150–5,881	6,000	Medicinal plants, bamboos, rice	Lowland tropical evergreen rainforests, cool temperate rainforests, pine–oak forests, subalpine scrub	GS (705 km²), WS (215 km²); important plant sites not protected	Slash-and-burn agriculture	Threatened; inadequate coverage of protected areas
Taninthayi (Tenasserim)	F	73,845	0–2,275	3,000	Timber trees (especially teak, rosewood), dyes, vegetables, fibers, medicinal plants	Wet evergreen dipterocarp forests, montane rain forests, bamboo scrub, mangroves	WS (49 km²), GS (139 km²), proposed NR (259 km²)	Logging, resin tapping, fuelwood cutting	Severely threatened in north; inadequate coverage of protected areas
SRI LANKA Knuckles	S	182	1,068–1906	>1,000	Timbers, medicinal plants, bamboos, fruit-tree relatives, spices, ornamentals	Lowland dry semievergreen, montane evergreen forests, grassland	No legal protection	Clearance for cardamom, settlements, and agriculture, mining, fuelwood cutting	Area below 1,200 m under threat
Peak Wilderness and Horton Plains	S	224/32	700–2,238/ 1,800–2,389	>1,000	Timbers, medicinal plants, bamboos, fruit-tree relatives, spices, ornamentals	Lowland, submontane, and montane wet evergreen forests, montane grassland	Sanctuary (Peak Wilderness), NP (Horton Plains)	Religious tourism, fuelwood and timber cutting, mining, fire, invasive grasses	Reasonably secure, but management plan required; some rich forests in upper regions should also be included in protected areas

Appendix 11.1 (continued)

Site	Type[a]	Area[b] (km²)	Altitude (m)	Flora[c] (N)	Examples of useful plants	Vegetation	Protected areas[d]	Threats[e]	Assessment[f]
Sinharaja	S	112	210–1,170	700 angiosp.	Timber trees, rattans, fruit-tree relatives, medicinal plants, ornamentals	Lowland and submontane wet tropical evergreen rainforests, grassland	National heritage wilderness area and WHS	Population pressure, encroachment of agriculture, overcollection of medicinal plants, potential hydroelectric scheme	Southern part still under threat; otherwise threats declining and reasonably secure
South East Asia (Malesia)									
BRUNEI DARUSSALAM									
Batu Apoi FoR, Ulu Temburong	S	488	50–1,850	3,000	Timber trees (especially dipterocarps, *Agathis*), medicinal plants, ornamentals	Lowland dipterocarp rainforests, lower and upper montane forests	FoR, planned NP	Logging, potential dam construction	At risk
INDONESIA (IRIAN JAYA)									
Arfak Mountains	F	2,200	100–3,100	3,000–4,000	Timber trees, rattans, fruit-tree relatives, ornamentals (especially rhododendrons)	Lowland, hill, and lower montane rainforest, grassland or heath communities, lake vegetation	NR (450 km²); proposal to extend to 653 km² as nature conservation area	Population pressure, resettlement schemes, agriculture and logging, road building	Lowlands threatened in particular

Appendix 11.1 (continued)

Site	Type[a]	Area[b] (km²)	Altitude (m)	Flora[c] (N)	Examples of useful plants	Vegetation	Protected areas[d]	Threats[e]	Assessment[f]
Gunung Lorentz	S	21,500	0–4,884	3,000–4,000	Fruits, vegetables, fibers, building materials	Lowland to montane rainforests, mangroves, bogs, swamps, heaths, grassland, alpine vegetation	NR; proposals for NP, BR, and WHS status	Mining, logging, petroleum exploitation, road building, tourism, colonization	Threatened; high-altitude vegetation vulnerable to trampling
Mamberamo-Pegunungan Jayawijaya	F	23,244	0–4,640	2,000–3000	Timber trees (especially southern beech, podocarps, conifers)	Lowland to montane rainforests, lowland swamp forests, mangroves	Proposed NP/WHS (14,425 km²), GR (8,000 km²), proposed GR (819 km²)	Petroleum exploitation, logging at lower altitudes	Lowlands particularly at risk
Waigeo	S	14,784	0–999		Timber trees, wild sugar cane	Lowland to lower montane rainforests, riverine forests, mangroves, limestone and ultra-mafic vegetation	NR (1,530 km²); marine reserve covers offshore islets and reefs	Potential nickel mining	At risk if mining goes ahead
INDONESIA (JAVA)									
Gede-Pangrango NP	S	150	1,000–3,019	>1,000	Timber trees, medicinal plants, ornamentals	Mostly montane and submontane rainforests, grass plains	NP (whole area), BR (140 km²)	Timber and fuelwood cutting, agricultural encroachment, visitor pressure, plant collecting	Encroachment around boundaries; at risk

Appendix 11.1 (continued)

Site	Type[a]	Area[b] (km²)	Altitude (m)	Flora[c] (N)	Examples of useful plants	Vegetation	Protected areas[d]	Threats[e]	Assessment[f]
INDONESIA (KALIMANTAN) Bukit Raya and Bukit Baka	S	7,705	100–2,278	2,000–4,000	Timber trees (especially dipterocarps), fruit trees, illipe nuts, rattans	Lowland tropical rainforests, swamp forests, lower and upper montane forests, ericaceous scrub	NP (1,811 km²)	Logging, road construction, shifting cultivation	Encroachment in west; at risk
Gunung Palung	S	900	0–1,160		Timber trees, fruit trees, ornamentals	Dipterocarp rainforests, montane forests, swamp forests, beach forests, mangroves	NP	Logging, shifting cultivation	Buffer zones needed; most of area safe, but some parts at risk
Sungai Kayan–Sungai Mentarang NR	S	29,000	100–2,556	2,000	Timber trees, fruit trees, gingers, rattans	Lowland and hill dipterocarp rainforests, montane forests, riverine, swamp and heath forests	NR (16,000 km²); proposed extensions 13,000 km²)	Logging, mining, shifting cultivation	Boundaries at risk
INDONESIA (SULAWESI) Dumoga-Bone NP	S	3,000	200–1,968		Timber trees, rattans	Tropical lowland semi-evergreen rainforests, riverine forests, montane forests, some limestone forests	NP	Overcollection of forest products, shifting cultivation, potential mining, road building	Boundaries threatened, but demarcation and zoning being implemented

Appendix 11.1 (continued)

Site	Type[a]	Area[b] (km²)	Altitude (m)	Flora[c] (N)	Examples of useful plants	Vegetation	Protected areas[d]	Threats[e]	Assessment[f]
Limestone flora	V		150–1,000		Sugar palms, fruit-tree relatives, plants for degraded land	Forests over limestone, scrub, lithophytic vegetation	c. 70% of all outcrops unprotected	Quarrying, firewood cutting, clearance for agriculture, fire	Threatened
Pegunungan Latimojong	S	580	1,000–3,455		Ornamentals	Lower and upper montane forests, hill forests, montane grassland, subalpine vegetation	PF; proposed NR	Clearance of lower slopes for agriculture	Lowlands threatened, but most of area not threatened
Ultramafic flora	V	12,000			Plants of potential value for rehabilitating degraded areas	Ultramafic facies of lowland forests, scrub, some montane vegetation	Mostly unprotected	Mostly intact, but agricultural development planned	Threatened
INDONESIA (SUMATRA)									
Gunung Leuser	S	>9,000	0–3,466	2,000–3,000	Timber trees (especially dipterocarps), fruit trees, medicinal plants, ornamentals	Lowland dipterocarp rainforests, montane and subalpine forests, freshwater swamp forests, marshes	NP (7,926 km²), BR (9,464 km²)	Encroachment of settlements, agriculture, illegal logging, overcollection of rattans	Threatened, especially in lowlands; inadequate funding
Kerinci-Seblat NP	S	1,484	200–3,805	2,000–3000	Fruit trees, timbers (e.g., dipterocarps, Agathis), medicinal plants, rattans	Lowland and hill dipterocarp rainforests, montane forests, montane swamp forests	NP	Encroachment of settlements, agriculture, illegal logging, overcollection of rattans	Threatened, especially in lowlands

Appendix 11.1 (continued)

Site	Type[a]	Area[b] (km²)	Altitude (m)	Flora[c] (N)	Examples of useful plants	Vegetation	Protected areas[d]	Threats[e]	Assessment[f]
Limestone flora	V	5,000	150–1,500	1,500–2,000	Ornamentals, fruit-tree relatives	Forests over limestone, scrub, lithophytic vegetation	Very few outcrops protected	Quarrying	Threatened
Tigapuluh Mountains	S	2,000	150–800	2,000–3,000	Timber trees	Tropical lowland evergreen and hill dipterocarp rainforests	None	Logging, conversion to forestry and rubber plantations, shifting cultivation	Eastern lowlands severely threatened
INDONESIA/ MALAYSIA									
Lanjak Entimau WS, Batang Ai NP, Gunung Bentuang dan Karimun	S	10,111	500–1,284		Timber trees (especially dipterocarps), illipe nuts, rattans, fruit-tree relatives	Lowland and hill evergreen rainforests, heath, swamp and montane forests	WS (1688 km²), proposed extensions (184 km²), NP (240 km²), NR (8,000 km²)	Agricultural encroachment, logging	Boundaries and lower slopes at risk
MALAYSIA (SABAH)									
East Sabah lowland and hill dipterocarp forests	V	20,000	0–1,298	5,000–6,000	Timber trees (seraya, keruing, kapur), rattans, fruit trees, ornamentals	Tropical evergreen lowland and hill dipterocarp rainforests	4,077 km² protected in conservation areas	Conversion to agriculture, human settlement, tree plantations, unsustainable logging	Threatened, some areas seriously threatened; protection of reserves needs strengthening
Kinabalu Park	S	753	150–4,101	4,500	Timber trees, ornamentals, (e.g., pitcher plants, orchids)	Mostly montane rainforests, some tropical lowland rainforests, ultramafic forests, alpine vegetation	SP	Clearance for cultivation, illegal logging, mining, tourism	Boundaries threatened

Appendix 11.1 (continued)

Site	Type[a]	Area[b] (km²)	Altitude (m)	Flora[c] (N)	Examples of useful plants	Vegetation	Protected areas[d]	Threats[e]	Assessment[f]
Northeast Borneo Ultramafic flora	V	3,500	0–3,000		Timber trees, potential ornamentals, nickel- and manganese-tolerant species	Ultramafic facies of tropical lowland evergreen rain-forests, lower and upper montane forests	Kinabalu SP only area of formal protec-tion; small areas in Danum Valley Conservation Area; Mt. Silam is a protected watershed	Clearance for cultivation and golf course, logging, fire, dam construction, road building	Inadequate coverage of protected areas; threatened
MALAYSIA (SARAWAK) Lambir Hills	S	69	30–467	1,500	Timber trees (including 69 dipterocarp spp.), mango and durian relatives, rattans	Lowland mixed dipterocarp forests, heath forests, scrub	NP	Logging, clearance for agriculture	Encroachment around boundaries; threatened
MALAYSIA/BRUNEI Gunung Mulu NP; Medalam PF; Labi Hills; Bukit Teraja; Ulu Ingei; Sungei Ingei	S	1,521	30–2,376	3,500	Timber trees (especially dipterocarps), fruit and nut trees, sago palms, rattans, medicinal plants	Lowland mixed dipterocarp to montane forests on sandstones, lime-stones, and shales; heath forests; peatswamp forests	Malaysia: NP, PF; Brunei: PF and conservation area within FoR	Logging around perimeter of Mulu NP, shifting cultiva-tion along some rivers, potential threat from road construction	Mostly safe at present, but would be at risk if proposed road went ahead; buffer zones to Mulu NP need to be implemented
Limestone flora of Borneo	V		0–1,710		Ornamentals (especially orchids, gesneriads, bego-nias, balsams, ferns)	Lowland to upper montane forests on limestone, litho-phytic vegetation	Some major areas in Sarawak in NP, few in Sabah in FoRs, proposed NPs in Kaliman-tan, other sites unprotected	Quarrying, fire, clearance of sur-rounding forests for agriculture, tourism	Some major areas safe, some relatively safe but unpro-tected, most others threatened

Appendix 11.1 (continued)

Site	Type[a]	Area[b] (km²)	Altitude (m)	Flora[c] (N)	Examples of useful plants	Vegetation	Protected areas[d]	Threats[e]	Assessment[f]
MALAYSIA (PENINSULAR MALAYSIA)									
Endau–Rompin State Parks	S	500	100–1,000		Timber trees (including dipterocarps), rattans, fruit trees, banana relatives, medicinal herbs	Mainly tropical lowland rainforests and hill dipterocarp forests, hill swamp forests	FoR; proposed state parks	Logging, clearance for development schemes, tourist facilities, commercial collection of ornamentals	Protection needs strengthening; at risk
Limestone flora	V	260	0–713	>1,300	Ornamentals (especially orchids, begonias, palms, gesneriads)	Limestone forests, scrub, lithophytic vegetation	A few outcrops protected in Taman Negara (NP); some occur in FoRs; some are temple reserves (no protection to flora)	Quarrying, mining, encroachment from agriculture, fire, tourism, plant collecting	Many outcrops severely threatened, some at risk, a few safe
Montane flora	V	2,180	810–2,188	>3,000	Ornamentals (especially orchids, pitcher plants, rhododendrons)	Lower and upper montane forests	Most peaks fall within FoRs, G. Tahan is in NP, Cameron Highlands is in WS, G. Kajang is in WR	Large-scale resort development, agriculture and horticulture, road building, plant collecting	Most areas outside NP severely threatened

Site	Type[a]	Area[b] (km²)	Altitude (m)	Flora[c] (N)	Examples of useful plants	Vegetation	Protected areas[d]	Threats[e]	Assessment[f]
Pulau Tioman	S	72	0–1,038	1,500	Ornamental plants (especially slipper orchids), *Rafflesia* used medicinally	Coastal forests, hill and upper montane forests, some mangroves	WR	Large-scale resort development, airstrip construction, overcollecting of orchids and *Rafflesia*	Threatened; protected status not enforced
Taman Negara	S	4,343	75–2,188	>3,000	Timber and fruit trees, rattans, ornamentals (e.g., orchids), potential medicinal plants	Lowland, hill and montane rainforests, "padang" vegetation, limestone and quartzite vegetation, riparian communities	NP	Logging, hydroelectric dams, lack of buffer zone, some tourist developments	Safe at present, but frequently threatened
Trengganu Hills	S	150	60–920	1,500	Timber trees, rattans, ornamentals	Lowland and hill rainforests	FoR, 3 small VJRs	Logging, land clearance for cultivation	At risk
PAPUA NEW GUINEA									
Bismarck Falls; Mt. Wilhelm; Mt. Otto; Schrader Range; Mt. Hellwig; Gahavisuka	S	9,754	250–4,499	5,000–6,000	Traditional food and medicinal plants, timbers, fibers, plants of cultural value	Lowland swamp and rainforests, montane forests, alpine vegetation	Proposed NP (Mt. Wilhelm), small provincial park (Gahavisuka); region proposed as WHS	Population pressure, logging, agriculture, coffee and cardamom plantations	Protected area coverage inadequate; at risk

Appendix 11.1 (continued)

Site	Type[a]	Area[b] (km²)	Altitude (m)	Flora[c] (N)	Examples of useful plants	Vegetation	Protected areas[d]	Threats[e]	Assessment[f]
Huon Peninsula (Mt. Bangeta; Rawlinson Range; Cromwell Ranges; Sialum Terraces)	S	3,415	0–4,120	4000–5000	Fruit trees, vegetables, fibers, potential timber species	Lowland tropical rainforests to subalpine forests, grassland, mangroves	No formal protection	Logging, road building	At risk
Menyamya; Aseki; Amungwiwa; Bowutu Mts.; Lasanga Island	F	6,695	0–3,278	1,500–3,000	Fruits, vegetables, fibers, building materials, potential timber species	Lowland rainforests to upper montane forests, ultramafic vegetation	NP (20 km²); local support for conservation area in Bowutu Mts.; several reserves needed throughout region	Logging, road building, local population pressure	Severely threatened in places, protected area coverage inadequate
Mt. Giluwe; Tari Gap; Doma Peaks	S	3,346	1,000–4,368	>3,000	Traditional food and medicinal plants, fibers, ornamentals	Montane and subalpine forests, grassland, alpine communities	Local reserves in Tari Gap area	Logging, road building, clearance for agricultural plantations, dieback of *Nothofagus*	Protected area coverage inadequate; at risk
Southern Fly Platform	F	18,644	0–100	>2,000	Edible palms, traditional food and medicinal plants	Savanna, monsoon forests, mangroves, lowland swamps	Wildlife management area (5,900 km²)	No major threats, some grazing pressure from introduced deer, potential threat from mining, some clearance for gardening	Reasonably safe, but protected area coverage inadequate

Appendix 11.1 (continued)

Site	Type[a]	Area[b] (km²)	Altitude (m)	Flora[c] (N)	Examples of useful plants	Vegetation	Protected areas[d]	Threats[e]	Assessment[f]
PHILIPPINES Batan Islands	S	209	0–1,008	>500	Timbers, fibers, medicinal plants, food plants	Lowland evergreen to mid-montane rainforests, grasslands, secondary vegetation	Proposed as protected landscape and 2 "critical watersheds" under National Integrated Protected Area System	Typhoons, clearance for grazing, crops, shifting cultivation, overcollection of forest products	Vulnerable, but growing conservation awareness among local people
Mt. Apo	S	769	500–2,954	>800	Ornamental (e.g., orchids, aroids, begonias), timber trees	Lowland rainforests (mostly cleared), montane forests, "elfin woodland," scrub, grasslands	NP	Construction of geothermal plant, clearance for agriculture, illegal logging, shifting cultivation	Severely threatened
Mt. Pulog	S	115	2,600–2,929	800	Timber tree provenances (especially pines), ornamentals	Montane forests, pine forests, grasslands	NP	Conversion of forest to vegetable and cut-flower gardens, fire	Threatened
Palanan Wilderness Area	S	2,168	0–1,672	1,500	Timber trees, rattans	Lowland and hill dipterocarp forests, lower montane forests, ultramafic and limestone forests	wilderness area; proposed as NP	Illegal logging, shifting cultivation, overcollection of forest products, potential large-scale logging	Mostly safe at present, but could become severely threatened

Appendix 11.1 (continued)

Site	Type[a]	Area[b] (km²)	Altitude (m)	Flora[c] (N)	Examples of useful plants	Vegetation	Protected areas[d]	Threats[e]	Assessment[f]
Palawan	F	14,896	0–2,085	>2,000	Timber trees, rattans, almaciga resin, fruit trees, orchids, nipa palms	Lowland evergreen dipterocarp and semi-deciduous forests, ultramafic and limestone forests, mangroves	BR (11,508 km²), NP (39 km²), various other protected areas covering 3.4% of land area	Logging, mining, shifting cultivation, tourism, overcollection of forest products	Inadequately protected; threatened
Sibuyan Island	S	445	0–2,052	700	Timber trees, almaciga resin, ornamentals	Lowland dipterocarp forests, montane forests, grasslands, mangroves	Proposed NR	Logging, slash-and-burn agriculture, fire, overexploitation of rattans	Threatened

[a] S = Site: the area is a discrete geographical unit, and the whole area needs to be conserved; F = Floristic province (often a wide area, or a CPD site covering a whole region): effective conservation of the flora of such areas often requires a network of reserves to be established, because in many cases it would be impractical to protect the entire province or region; V = Vegetation type: as with floristic provinces, effective conservation often requires representative samples to be protected.

[b] Usually given to the nearest 1 km² for individual sites; for large regions, sometimes an estimate to the nearest 100 km² or 1,000 km² is given.

[c] Unless otherwise stated, numbers refer to indigenous vascular plant species, or an estimate based on current botanical knowledge of that site (or similar sites), usually to the nearest 100 species, or to nearest 1,000 species for some large tropical sites. *Taxa* refers to the number of species, subspecies, and varieties. *Angiosp.* = angiosperms.

[d] Categories of protected areas are given where a CPD site is fully or partially protected. In most cases, the area of the protected site is given after the category, or a percentage figure is given for the part of the site that is protected. If a site is fully protected, the entry will show just the category of protection. BR = Biosphere Reserve; FoR = Forest Reserve; GR = Game Reserve; GS = Game Sanctuary; NM = Nature Monument; NP = National Park; PF = Protection Forest or Protected Forest; RF = Reserved Forest; SP = State Park; VJR = Virgin Jungle Reserve; WHS = World Heritage Site; WR = Wildlife Reserve; and WS = Wildlife Sanctuary.

[e] Main threats are listed in order of importance.

[f] Threats and conservation status of CPD sites are analyzed in tables 11.2 and 11.3.

[g] The exact number of plant species present or recorded for this site.

Campbell, D. G. and H. D. Hammond, editors. 1989. *Floristic Inventory in Tropical Countries: The Status of Plant Systematics, Collections, and Vegetation.* New York Botanical Garden, New York.

Davis, S. D., V. H. Heywood, and A. C. Hamilton, editors. *Centres of Plant Diversity: A Guide and Strategy for Their Conservation.* 1994 Vol. 1: *Europe, Africa, South West Asia and the Middle East.* IUCN Publications Unit, Cambridge, England.

Dinerstein, E., and E. D. Wikramanayake. 1993. Beyond "hotspots": How to prioritize investments to conserve biodiversity in the Indo-Pacific region. *Cons. Biol.* 7: 530–565.

Dinerstein, E., E. D. Wikramanayake, and M. Forney. This volume. Conserving the reservoirs and remnants of tropical moist forest in the Indo-Pacific region.

Hawkes, J. G. 1983. *The Diversity of Crop Plants.* Harvard University Press, Cambridge.

Heywood, V. H. 1992. Conservation of germplasm of wild species. In *Conservation of Biodiversity for Sustainable Development.* O. T. Sandlund, K. Hindar, and A. H. D. Brown, editors. Scandinavian University Press, Oslo, Norway.

———. 1993. The measurement of biodiversity and the politics of implementation. In *Systematics and Conservation Evaluation.* P. L. Forey, C. J. Humphries, and R. I. Vane-Wright, editors. Oxford University Press, Oxford.

ICBP (International Council for Bird Preservation). 1992. *Putting Biodiversity on the Map: Priority Areas for Global Conservation.* ICBP, Cambridge, England.

IUCN (International Union for the Conservation of Nature and Natural Resources). 1980. *World Conservation Strategy: Living Resource Conservation for Sustainable Development.* IUCN, UNEP, and WWF, Gland, Switzerland.

Mittermeier, R. A., and T. B. Werner. 1988. Wealth of plants and animals unites "megadiversity" countries. *Tropicus* 4: 4–5.

Myers, N. 1988. Threatened biotas: "Hotspots" in tropical forests. *Environmentalist* 8: 187–208.

———. 1989. *Deforestation Rates in Tropical Forests and Their Climatic Implications.* Friends of the Earth, London.

———. 1990. The biodiversity challenge: Expanded hot-spots analysis. *Environmentalist* 10: 243–256.

Proctor, J. This volume. Rainforests and their soils.

Raven, P. H. 1987. The scope of the plant conservation problem world-wide. In *Botanic Gardens and the World Conservation Strategy: Proceedings of an International Conference, Las Palmas de Gran Canaria.* D. Bramwell, O. Hamann, V. Heywood, and H. Synge, editors. Academic Press, London.

———. 1988. Our diminishing tropical forests. In *Biodiversity.* E. O. Wilson, editor. National Academy Press, Washington, D.C.

Stattersfield, A. J., M. J. Crosby, A. J. Long, and D. C. Wege. In prep. *Global Directory of Endemic Bird Areas.* BirdLife International, Cambridge, England.

Vane-Wright, R. I., C. J. Humphries, and P. H. Williams. 1991. What to protect and the agony of choice. *Biol. Cons.* 55: 235–254.

Whitmore, T. C., and J. A. Sayer, editors. 1992. *Tropical Deforestation and Species Extinction.* IUCN, Gland, Switzerland, and Cambridge, England, and Chapman and Hall, London.

Wilson, E. O. 1988. The current state of biological diversity. In *Biodiversity.* E. O. Wilson, editor. National Academy Press, Washington, D.C.

12

THE ROLE OF TOTALLY PROTECTED AREAS IN PRESERVING BIOLOGICAL DIVERSITY IN SARAWAK

Abang Haji Kassim bin Abang Morshidi and
Melvin Terry Gumal

Sarawak is the largest of the thirteen states in Malaysia, occupying a landmass of 12.3 million ha between latitudes 0°50' and 5°10' north and longitudes 109°35' and 115°40' east. The area under natural forest cover in Sarawak is about 8.7 million ha (Sarawak Forest Department 1991). Of this, about 6 million ha are designated and managed as Permanent Forest Estates (PFEs), and 288,806 ha are Totally Protected Areas (TPAs). TPAs, which comprise about 3.3% of the state's natural forest cover, fall into three categories: national parks (the nine in Sarawak cover 113,955 ha), wildlife sanctuaries (there are three, totaling 174,851.4 ha), and nature reserves (fig. 12.1 and table 12.1). National parks are open to the

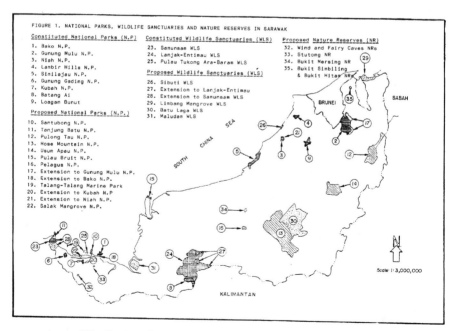

FIGURE 12.1 Distribution of actual and proposed national parks, wildlife sanctuaries, and nature reserves in Sarawak.

Table 12.1   Areas in the Totally Protected Area system by major forest type

	MDF[a] (ha)	Peatswamp (ha)	Mangrove (ha)	Heath (shrub) (ha)	Heath (pole) (ha)
National parks	55,876	8,578	160	3,218	32,631
	285,775		*10,249*	*11,428*	*16,571*
Total	**341,651**	**8,578**	**10,409**	**14,646**	**49,202**
Wildlife sanctuaries	139,743	—	212	3,856	29,196
	154,150	*9,710*	*2,324*	*14,243*	*22,119*
Total	**293,893**	**9,710**	**2,536**	**18,099**	**51,315**
Nature reserves	—	—	—	—	—
	1,147	—	—	—	—

Note: Italicized numbers indicate proposed areas.

[a] Mixed dipterocarp forest.

public for recreation, but public access is limited in the wildlife sanctuaries, which are strictly for conservation and research. Nature reserves are the same as national parks, but smaller; they are less than 1,000 ha (*National Parks* 1990). Under the proposed system, the percentage of natural forest cover included in the TPAs would increase to around 11.5% of Sarawak's forest area (over 8% of Sarawak's land area).

As a signatory to the Convention on Biological Diversity, Malaysia, and Sarawak in particular, is aware of the value of biological diversity and is concerned that this diversity is being significantly reduced by certain human activities. Biodiversity is defined as "the variability among living organisms from all sources including, among other things, terrestrial, marine and other aquatic ecosystems and the ecological complexes of which they are part; this includes diversity within species, between species and of ecosystems" (UNEP 1992). This chapter focuses primarily on ecosystem and species biodiversity, although considerations of genetic diversity are also included. A fundamental requirement for the conservation of biological diversity is the in situ conservation of ecosystems, which Sarawak promotes through its TPA system.

The idea of TPAs in Sarawak began with the conceptualization of national parks in the 1950s. The National Parks Ordinance was passed and a Board of Trustees formed to administer the national parks in 1956 (Abang Morshidi 1977). In 1957, the first national park, Bako National Park, was legally constituted. In 1973, the Conservator of Forests replaced the Board of Trustees as administrator of national parks, and the Forest Department was charged with the responsibility of enforcing the National Parks Ordinance (Amendment 1973). The Wildlife Protection Ordinance was drawn up in 1958. Samunsam Wildlife Sanctuary was the first wildlife sanctuary established under this ordinance in 1979. Both the National Parks Ordinance and the Wildlife Protection Ordinance have since been amended to create a new category of totally protected areas—nature reserves, formally established in 1990 under the amended National Parks Ordinance.

THE ROLES OF TPAS IN SARAWAK

Sarawak's TPAs are managed to generate sustainable benefits for society. For example, parks provide an important source of income for rural people by supplying jobs related to recreation and harvest of renewable resources. Local communities use forest resources to fulfill domestic needs. The areas also provide more subtle benefits, such as protected groundwater supplies and topsoil, and a moderated climate. Furthermore, TPAs permit the conservation of habitats and endangered species while supporting biological and ecological research and education.

The forests' resources provide income in several sectors of the economy. Tourism is particularly lucrative; in 1991, for example, 188,135 foreign tourists contributed $MR180.6 million (1 Malaysian ringgit equals 0.37 U.S. dollars) to Sarawak's economy. Of these tourists, 23,228 visited one or more of the four parks—Bako, Niah, Lambir Hills, and Gunung Mulu. This figure represents 12% of all visitors entering the state (Tourism Development Corp. Malaysia, personal communication, 1992) and affects roughly $MR20 million of the tourist economy. Chung (1987) calculated that the recreational value (through travel cost method) of three national parks (Bako, Lambir Hills, and Niah) in any one year is several million ringgit.

Local communities depend on the forests for resources for domestic use (such as plants, fruits, fishes, and wild meat), which residents are allowed to collect within a TPA. Many local people have collected materials from the forest for generations and fully expect to continue gathering supplies from all forested lands near their villages. Totally protected areas become important reserves of valued forest products as these products are depleted by overcollection, logging, and farming outside of the TPAs. The right of access to forests and forest products must be legally determined before the TPA is constituted. This process is frequently lengthy and difficult because written documentation and surveys are usually not available, and villages and people often move from place to place over time.

The TPAs also protect other vital resources, such as water catchments. Services provided by TPAs may have no immediate monetary value, but they are of tremendous importance to local and regional populations. Streams in Lambir Hills National Park and Kubah National Park, for example, supply about 38 million liters of water daily to Miri and Kuching, respectively (Ngui 1991). If the state government had to develop alternative water supplies for these municipal areas, the cost would be enormous.

Further, TPAs are venues for environmental education. Together with other institutions, both local and overseas, the Sarawak Forest Department is using the TPAs to train teachers in environmental education. The Forest Department is also running conservation education programs with the goal of increasing local participation in conservation and sustainable use of resources. Local communities benefit from the sustainable use of protected areas and can contribute to maintaining biological diversity (McNeely et al. 1990). At Batang Ai National Park, for instance, a pilot conservation education project was initiated with local rural communities to increase their involvement in park management. The project aims to cooperate with local communities to maintain or enhance the biodiversity of the national park and the surrounding areas (Gumal and Tan 1992). Local people have started to keep track of hunters in the national park, and they prevent outsiders without hunting privileges from entering (E. L. Bennett, personal communication, 1992). The key to developing such cooperation with local communities is the ability to demonstrate that national parks have economic benefits that compensate for the restrictions park regulations impose on the use of forest resources.

BIOLOGICAL DIVERSITY

Sarawak intends to represent all the natural ecosystems, communities, and species in its TPA system (Kavanagh, Rahim, and Hails 1989; Primack 1991). Of particular importance is the conservation of endangered and threatened species. Almost all the flora and fauna listed as totally protected (endangered) and protected (threatened) are found inside the existing or proposed TPAs (appendix 12.1). The only exceptions are dugongs (*Dugong dugon*) and Sumatran rhinoceroses (*Dicerorhinus sumatrensis*). Dugongs are sometimes still found off the coast of Lawas (F. Gombek, personal communication, 1992). The range of the Sumatran rhinoceros is unknown, but its footprints have been found within the Permanent Forest Estate, outside the borders of the proposed Pulong Tau National Park (D. Labang, personal communication, 1992). The Forest Department is monitoring the movements of this species to determine its territory.

Protecting species and ecosystems can best be done by protecting habitats (McNeely et al. 1990). Sarawak supports diverse forest habitats, which vary in plant species composition, and exceptionally high local levels of both plant diversity and endemism (Ashton 1989; see table 12.1 and appendix 12.1). In particular, Sarawak is a center of species richness for timber trees in the family Dipterocarpaceae. All major habitat types in Sarawak are represented in the proposed TPA system, though some, such as mangroves and peatswamp, are currently underrepresented. In addition, certain vegetation types vary considerably. The mixed dipterocarp forest in particular varies widely in species composition and forest structure. Once the proposed areas are included in the TPA system, protection for these major habitats will grow: Sarawak Mangroves National Park will increase by 8,682 ha, Limbang Mangroves Wildlife Sanctuary by 4,500 ha, and Maludam Wildlife Sanctuary (peatswamp) by 8,700 ha (Sarawak Forest Department 1991).

The Sarawak Forest Department also carries out ex situ conservation of rare and threatened plants in the Botany Research Centre, Kuching. It has collected and planted more than 3,000 local plants. The collection emphasizes groups such as orchids, pitcher plants, palms, wild fruits, local medicinal plants, economically important species, and attractive herbaceous plants (Choon and Mamit 1990). The Sarawak Forest Department does not carry out ex situ conservation of wildlife, but its Wildlife Rehabilitation Centre in Kuching rehabilitates confiscated wild animals to train them to adapt to the wild.

DISCUSSION

The studies of MacKinnon and MacKinnon (1986) showed that the protected area system in Borneo (including Sarawak) is fairly good and could be very good if half of the proposed additions are realized. Levels of protection for the areas could be improved but are generally adequate to cope with threats of illegal logging, hunting, shifting cultivation, and illegal settlements. To conserve biodiversity effectively and sufficiently in Sarawak, the Forest Department must address a number of specific problems. First, Sarawak, like most tropical countries, has many endemic species and a high level of diversity in its plants and animals. Some of these species may be "missed" by the parks—unique species or ecosystems may not be incorporated in the TPA system. Comprehensive studies by Kavanagh (1985); Ngui and Chai (1985); Kavanagh, Rahim, and Hails (1989); and Ngui (1991) demonstrate that although many areas of high conservation value fall within the current and proposed tpa system, some isolated areas outside the proposed system may be of biological importance. In the continual upgrading of the TPA system, these areas would be incorporated where necessary. An example is Loagan Bunut National Park, which provides extra protection to peatswamp forests as compared to 1985.

Second, effective conservation of biological diversity must be based on accurate information. The state of knowledge about most species and ecosystems is inadequate; the only detailed wildlife research has concentrated on proboscis monkeys (Bennett 1986, 1988, 1989; Bennett and Sebastian 1988), hornbills (Kemp and Kemp 1975), use of different habitats at Samunsam by primate communities (Rajanathan 1992), and specific problems for conservation of animal biodiversity, notably hunting (Caldecott 1988) and selective timber extraction (Zainuddin, in prep.). Short surveys have been undertaken on orangutans, clouded leopards, Sumatran rhinoceroses, terns, migrant birds, and swiftlets, but detailed ecological studies must still be carried out for these and many other animal species. Botanical studies are equally lacking: although much information about Sarawak's timber trees exists, the information on the conservation status of native wild plants is generally no better than that available for large animals. For certain groups, such as palms (Pearce 1989) and orchids (Lamb 1991), however, adequate information is available to indicate priority areas for inclusion in the TPA system.

The Forest Department recognizes that it does not have sufficient data for the detailed management of wildlife. This problem stems from the lack of trained personnel and staff in this field. The department will continue to collaborate with such international organizations as the World Wide Fund for Nature–Malaysia, Wildlife Conservation International, and local institutions on wildlife research projects. For example, the department is involved in wildlife and plant research conducted on an ecological plot in Lambir Hills National Park in collaboration with the various Japanese universities and Harvard University. The department's study of all animals and plants in Lanjak-Entimau Wildlife Sanctuary aims to develop a management plan for the sanctuary, in collaboration with International Tropical Timber Organization (ITTO). By creating collaborative ventures with outside organizations, the Forest Department can collect the information it needs, despite the lack of personnel.

Third, a reserve should be large enough to contain minimum viable populations of the endangered species. If a single large reserve is not feasible, reserves should be established in several sites (Soulé 1988). These sites should have uncorrelated environments, so that major environmental perturbations do not occur simultaneously, or with the same severity in each site (Soulé 1988). Most nature reserves worldwide, including Biosphere Reserves, are under 1 million ha

(Frankel and Soulé 1981). Sarawak's TPAs vary in size from 1.4 ha (Pulau Tukong Ara-Banun Wildlife Sanctuary) to 168,758 ha (Lanjak-Entimau Wildlife Sanctuary). The reasons for constituting all the TPAs are shown in appendix 12.1. Each reserve is intended to support the full complement of its native mammalian fauna and does so independently of all other reserves and surrounding nonreserve areas without active management.

To conserve large mammalian species effectively with a high probability (95%) of persistence for even a century, Belovsky (1988) estimated that reserves on the order of 10–100 million ha are needed. This is unattainable in most countries, including Sarawak, because the natural forest area in Sarawak covers only 8.7 million ha. On the entire island of Borneo, however, such large areas of permanent forest cover are potentially attainable, because three countries, including six states and provinces, are represented. This point highlights the potential value of transfrontier protected areas. The goal of reserving a few large TPAs, however, should not detract from the need for a larger number of small TPAs chosen to preserve specific areas of high biological diversity (fig. 12.2). Most living species can probably survive in the long term in relatively small areas.

FIGURE 12.2 Many species have highly restricted distributions and need to be protected within designated nature reserves. Shown here is the male inflorescence of *Artocapus annulatus*, a fruit tree found only in a restricted area of limestone hills south of Kuching, Sarawak. This species is related to such major fruit-tree species as chempedak (*Artocarpus integer*), jackfruit, and breadfruit.

The Forest Department is investigating whether the current and proposed TPAs are large enough to sustain viable populations of endangered plant and animal species. For example, areas of roughly 13,700 ha and 14,600 ha are needed to support viable populations of gibbons and langurs, respectively, taking into account their average densities in Sarawak's forests (Bennett, personal communication, 1992). These estimates are based on an effective breeding population of 1,000 and assume that the area is lightly hunted. Sarawak's TPA system contains a few areas large enough to support viable populations of these animals: Gunung Mulu and Batang Ai National Parks, and Lanjak-Entimau Wildlife Sanctuary.

The lack of sufficiently large TPAs to preserve viable populations of many animals has prompted the government to plan several new reserves, including Pulong Tau, Usun Apau, and Hose Mountain National Parks and Batu Laga Wildlife Sanctuary. There are also proposals to expand the boundaries of the existing TPAs. In constituting its new TPAs, the Forest Department has taken into account several principles of habitat management. First, larger areas of habitat contain more species and have larger population sizes than do smaller areas, so a few large TPAs protect more species in the long run than many small ones (Soulé and Simberloff 1986). Second, small areas are more prone to extinctions from random environmental fluctuations (Goodman 1987). Finally, large protected areas have substantially less perimeter to patrol than many small reserves of the same total area (Ayres, Bodmer, and Mittermeier 1990).

The Forest Department is also reviewing its Permanent Forest Estate (PFE) management system in Sarawak. The PFE includes areas where timber harvesting is prohibited, such as slopes greater than 35° (A. Dimin, personal communication, 1992). These areas, known as protection forests, totaled about 580,000 ha in 1991. Along with protection forests, logged-over and disturbed forests are components of the PFE that are important for biodiversity conservation, although their biodiversity does not equal that of the pristine forests. A network of interconnecting PFEs is planned to coincide with the current TPAs to preserve representatives of all species and habitats. In this way, overall biological diversity can be enhanced by integrating TPAs into management plans at the landscape level. Forests in the PFE can provide corridors, allowing the dispersal of animals among TPAs. These forests can also act as buffer zones around the TPAs and provide additional resources to species residing in TPAs. The TPAs in turn can serve as focal points of recolonization, providing the dispersing animals and seeds needed to reestablish the original biological community on logged forest lands.

Minimum Viable Population for Endangered Animals

There is no universally valid number of animals below which an endangered species is not viable. Generally, for a population to stand a good chance of survival for at least a hundred years, its numbers should be in the low thousands (Soulé 1988). Each situation is unique, however, and the number, density, and distribution of a minimum viable population depend on the acceptable level of risk (Schonewald-Cox 1983). Acceptable risk levels could vary, for example, between 50% probability of persistence for 100 years and 99% probability of persistence for 1,000 years. Accurate population numbers are available only for some primates, such as proboscis monkeys in Samunsam Wildlife Sanctuary (E. L. Bennett, personal communication). About 160 monkeys inhabit the sanctuary (Bennett and Sebastian 1988), so according to the rough guide of minimum viable population, the population of proboscis monkeys in Samunsam is probably not viable without active management and enhancement. The department is embarking on more wildlife research to help in managing such endangered species.

Land Use

The Forest Department is under constant pressure to use TPAs for agriculture, aquaculture, urban development, and logging. Native people are also claiming more land as their property under customary law, which recognizes prior land use. Almost all conservation agencies throughout the world face this problem. No simple recipe determines how biological resources in each locality can best be conserved and how land should best be used to achieve the objectives of conservation. Ecological, social, political, economic, and technological factors all enter into the decisions, and each factor changes over time. Because the factors are interrelated, a change in one can have effects—sometimes unpredictable—on all the others (McNeely et al. 1990). The Forest Department recognizes the demands of land users, but it is actively carrying out work to establish all the proposed TPAs.

Hunting is a particular concern in TPAs. Almost any animal larger than mouthful-sized is liable to be captured and eaten in Sarawak (Caldecott 1988). In some TPAs (such as Gunung Mulu and Batang Ai National Parks), local populations are given privileges to hunt certain unprotected species. The Forest Department is attempting to reduce outsiders' hunting for sport, however, by carrying out patrolling, enforcement, publicity, and conservation education programs. In Batang Ai National Park, local people are helping to stop sport hunters from entering the park. With good conservation education and public relations, hunting of endangered species by locals with privileged access to the area can also be reduced considerably.

CONCLUSION

The TPA system in Sarawak faces various problems, including land disputes, hunting, and uncertainty about the appropriate size of each TPA. With limited resources, the Forest Department has developed methods to increase the effectiveness of the current reserves and is seeking to add new ones. Some TPAs, for example, may be too small, but the Forest Department is developing corridors within the PFE interconnecting TPAs to increase the area used by threatened species. Such methods will complement the TPAs in conserving biodiversity; however, destructive activities, such as sport hunting, need to be controlled in these areas. This problem and others could be better addressed with more staff and trained personnel. The Forest Department is restructuring to increase staff actively involved in TPA management.

All major habitats will be protected under the TPA system. The system will be continually upgraded, with the objective of representing all the local ecosystems, communities, and species. Current gaps will be plugged later. In addition, some research projects on plant and animal species are being carried out. Although detailed information on many important species is still lacking, the Forest Department plans more research in these areas. Local participation in the conservation education and management programs looks promising, but these programs need careful monitoring to ensure that they are achieving their objectives.

Appendix 12.1 National Parks (NPs) and Wildlife Sanctuaries (WSs) in Sarawak

Protected area	Some endangered animals[a]	Habitat types[b]	Reasons for constitution
Established National Parks			
Bako N.P.	*Nasalis larvatis, Presbytis cristata, Nycticebus coucang, Tarsius bancanus, Anthracoceros coronatus, Haliaeetus leucogaster*	MDF (356); cliff (91); beach (37); mangrove (160); heath (pole: 1,640; shrub: 321); riverine (73); peatswamp; secondary (51)	Great diversity of flora and fauna; recreation
Batang Ai NP	*Pongo pygmaeus, Hylobates muelleri, Presbytis frontata, Presbytis rubicunda, Rattufa affinis, Tarsius argus, Argusianus tarsus, Rhyticeros undulatus, Anorrhimus galeritus, Rhinoplax vigil, Buceros rhinoceros, Anthracoceros malayanus, Lophura bulweri, Neofelis nebulosa*	MDF (24,040); riverine; secondary forest; heath (pole)	Viable orangutan population, recreation, watershed for hydro dam, increased protection for Lanjak-Entimau WS
Ganung Gading NP	*Rattufa affinis, Buceros rhinoceros, Argusianus argus*	MDF (3,540); heath (pole: 520; shrub: 136)	Rare plant (*Rafflesia tuarmundae*), watershed, recreation
Gunung Mulu NP	*Hylobates muelleri, Presbytis rubicunda, Tarsius bancanus, Cheiromeles torquatus, Lophura bulweri, Berenicornis comatus, Anorrhimus galeritus, Rhyticeros corrugatus, Rhyticeros undulatus, Anthracoceros malayanus, Anthracoceros coronatus, Buceros rhinoceros, Rhinoplax vigil, Rattufa affinis, Neofelis nebulosa, Tadarida plicata, Macbaeramphus alcinus*	MDF (13,200); heath (pole: 26,850; shrub: 120); montane (6,490); freshwater swamp (4,960); limestone forest; alluvial forest	Spectacular cave system; cultural, geological, and archaeological importance; diverse flora and fauna; recreation; large bat populations
Kubah NP	*Anthracoceros malayanus, Argusianus argus, Rattufa affinis*	MDF (1,139); heath (pole: 1,011; shrub: 80)	Water catchment area, recreational site, palm diversity; scenic beauty

Appendix 12.1 (continued)

Protected area	Some endangered animals[a]	Habitat types[b]	Reasons for constitution
Lambir Hills NP	Presbytis melalophos, Hylobates muelleri, Anthracoceros malayanus, Anorrhinus galeritus, Rhyticeros corrugatus, Rhyticeros undulatus, Buceros rhinoceros, Rhinoplax vigil, Haliaeetus leucogaster, Argusianus argus	MDF (6,952); heath (pole)	Great plant diversity, waterfalls for recreation, scenic beauty
Loagan Bunut NP	Bubulcus coromandus, Ciconia stormi, Callagur borneoensis, Tomistoma schlegelii, Anthracoceros malayanus, Anthracoceros coronatus, Hylobates muelleri, Ratufa affinis, Haliaeetus leucogaster, Felis planiceps	Water (735); peatswamp (7,949); MDF (78); heath (pole: 1,483)	Peatswamp habitat, flora and fauna of lake and forest, protection of local fishing tradition
Niah NP	Cyrtodactylus cavernicolus, Lanthanotus borneoensis, Presbytis cristata, Presbytis rubicunda	MDF (999); limestone forest (1,165); peatswamp forest (163); heath (pole: 813); alluvial forest	Archaeological and historical importance, swiftlet and bat population, cave complex
Similajau NP	Anthracoceros malayanus, Anthracoceros coronatus, Buceros rhinoceros, Haliaeetus leucogaster, Presbytis melalophos chrysomeles, Presbytis cristata, Hylobates muelleri, Chelonia mydas, Eretmochelys imbricata	MDF (3,960); peatswamp (110); heath (pole: 2,730; shrub: 265); riverine forest; beach forest	Previously important nesting ground for turtles, golden beaches, recreation, aesthetic beauty
Proposed National Parks[c]			
Hose Mountain NP	Neofelis nebulosa, Rhinoplax vigil, Buceros rhinoceros, Rhyticeros corrugatus, Rhyticeros undulatus, Anthracoceros malayanus, Anorrhinus galeritus, Argusianus argus, Hylobates muelleri, Presbytis rubicunda, Presbytis frontata, Presbytis bosei	Heath (pole: 16,361; shrub: 11,199); MDF (166,963); montane forest (cloud cover hampering aerial interpretation; 37,900)	Great biodiversity, watershed

Appendix 12.1 (continued)

Protected area	Some endangered animals[a]	Habitat types[b]	Reasons for constitution
Pelagus NP	*Didrondra aspera, Upuna borneoensis*	MDF (1,754); Heath (pole: 162; shrub); harvested; riparian forest	Rapids, recreation, aesthetic value, endemic plant species
Pulau Bruit NP	*Tringa totanus, Calidris ferruginea Limicla falcinellus*	Coastal mudflat; mangrove forest (1,567)	Important migrant bird habitat, mangrove and fisheries conservation, recreation
Pulong Tau NP	*Neofelis nebulosa, Felis planiceps, Ducula badia, Presbytis, rubicunda, Presbytis bosei, Hylobates muelleri, Ratufa affinis, Anorrhinus galeritus, Rhinoplax vigil, Buceros rhinoceros, Anthracoceros malayanus, Argusianus argus, Lophura bulweri*	MDF (63,700); heath (pole); montane forest	Great diversity for flora and fauna, watershed area, protection of habitat for wild pigs that are a major food supply for rural people, recreation
Sarawak Mangroves NP	*Nasalis larvatus, Presbytis larvatus, cristata,* migrant birds	Mangrove (6,310); heath (pole: 1,697); secondary (675)	Populations of *N. larvatus, P. cristata,* and mangrove habitats; essential support for fisheries
Santubong NP	*Anthracoceros malayanus, Ichthyophage ichthyaetus, Haliaeetus leucogaster, Hyoticebus coucang, Tarsius bancanus*	MDF (2,720); heath (pole)	Archaeological importance, watershed protection, environmental and scenic value, recreation, cultural, legendary and historical importance
Talang-Satang NP	*Eretmochelys imbricata, Chelonia mydas, Caretta caretta, Lepidochelys olivacea, Ducula bicolor, Hydrozoa* spp., *Anthozoa* spp., *Haliaeetus leucogaster, Ichthyophage ichthyaetus, Cetaces* spp.	Coral reefs, rocky vegetation	Turtle nesting area
Tanjong Datu NP	*Hylobates muelleri, Presbytis cristata, Presbytis melalophos, chrysomelas, Ratufa affinis, Rhinoplax vigil, Buceros rhinoceros, Anthracoceros malayanus, Anthracoceros coronatus, Chelonia mydas, Eretmochelys imbricata*	MDF (878); heath (pole: 210; shrub: 67); rocky coastline; beach forest	Habitat conservation for endangered species, recreation, environmental watershed, scenic protection

Appendix 12.1 (continued)

Protected area	Some endangered animals[a]	Habitat types[b]	Reasons for constitution
Usun Apau NP	*Hylobates muelleri, Lophura bulweri, Ratufa affinis, Presbytis rubicunda, Argusianus argus, Anorrhinus galerius, Rhinoplax vigil, Buceros rhinoceros, Anthracoceros malayanus*	MDF (49,760); heath (pole); montane forest	Watershed, waterfall complex, great biodiversity, archaeological and cultural importance
Established Wildlife Sanctuaries			
Lanjak Entimau WS	*Pongo pygmaeus, Hylobates muelleri, Presbytis rubicunda, Presbytis frontata, Buceros rhinoceros, Berenicornis comatus, Anorrhinus galeritus, Rhyticeros corrugatus Rhyticeros undulatus, Anthracoceros malayanus, Anthracoceros coronatus, Rhinoplax vigil, Lophura bulweri, Ratufa affinis, Neofelis nebulosa*	MDF (137,429); heath (pole: 25,641; shrub: 3,856; photo gap (1,832)	Great biodiversity, orangutan habitat, watershed
Pulau Tukong Ara-Banun WS	*Sterna sumatriana, Sterna anaetheta*	Rocky islet	Nesting grounds for endangered birds
Samunsam WS	*Hylobates muelleri, Nasalis larvatus, Presbytis cristata, Presbytis melalophos chrysomelas, Buceros rhinoceros, Berenicornis comatus, Anorrhinus galeritus, Rhyticeros corrugatus, Rhyticeros undulatus, Anthracoceros malayanus, Anthracoceros coronatus, Rhinoplax vigil, Ratufa affinis, Neofelis nebulosa, Argusianus argus, Pandion haliatus, Ichthyophage ichthyaetus*	MDF (2,314); mangrove (212); heath (pole: 3,555); riverine; heath (shrub); secondary forests	Potentially viable population of *Nasalis larvatus*

Appendix 12.1 (continued)

Protected area	Some endangered animals[a]	Habitat types[b]	Reasons for constitution
Proposed Wildlife Sanctuaries			
Batu Laga WS	*Buceros rhinoceros, Berenicornis comatus, Anorrhinus galeritus, Rhyticeros corrugatus, Rhyticeros undulatus, Anthracoceros malayanus, Anthracoceros coronatus, Rhinoplax vigil, Ratufa affinis, Neofelis nebulosa*	MDF (137,665); heath (pole: 8,447); heath (pole: 9,472); photo gap (42,577)	Great biological diversity
Lanjak Entimau WS extension	*Pongo pygmaeus*	MDF (13,899); heath (pole: 40; shrub: 4,475)	To increase the habitats for orangutans and reduce the sanctuary's edge effects
Lawas-Limbang Mangrove WS	*Nasalis larvatus, Haliaeetus leucogaster, Ichthyophage ichthyaetus, Presbytis cristata*	Mangrove (4,500)	Habitat for *Nasalis larvatus*, vital support for fisheries
Maludam WS	*Nasalis larvatus, Presbytis melalophos cruciger, Ratufa affinis, Ciconia stormi, Anorrhinus galeritus, Buceros rhinoceros, Anthracoceros malayanus, Anthracoceros coronatus*	Peatswamp (8,700)	Population of *Nasalis larvatus*, peatswamp habitat
Samunsam WS extension	*Nasalis larvatus, Presbytis cristata, Hylobates muelleri, Presbytis melalophos chrysomales*	MDF (2,586); heath (pole: 11,935; shrub: 4,475); mangrove; mudflat	To include *Nasalis larvatus* and *Presbytis* spp. home and population range
Sibuti WS	Diverse bird population	Mangrove (196); peatswamp (1,010); Heath (pole: 7)	Conservation of habitats for diverse bird population

Sources: Anderson, Jermy, and Cranbrook 1982; Bennett, personal communication; Bennett and Walsh 1988; Good 1991; Ngui 1991; NPWO, n.d., 1987, 1992a, 1992b; and Watson 1985.

Note: MDF = mixed dipterocarp forest.

[a]The lists of endangered animals are not exhaustive.

[b]Numbers in parentheses indicate the area (in ha) of each habitat type.

[c]The proposed national park extension areas have basically the same habitat areas as the existing national parks. The Forest Department has sought extensions to increase the habitat areas of endangered wildlife.

REFERENCES

Abang Morshidi, A.H.K. 1977. The development of national parks in Sarawak. *Malay. For.* 40: 138–143.

Anderson, J. A. R., A. C. Jermy, and Cranbrook, Earl of. 1982. *Gunung Mulu National Park: A Management and Development Plan.* Royal Geographical Society, London.

Ashton, P. S. 1989. Sundaland. In *Floristic Inventory of Tropical Countries: The Status of Plant Systematics, Collections, and Vegetation, plus Recommendations for the Future.* D. G. Campbell and H. D. Hammond, editors. New York Botanical Garden with WWF, New York.

Ayres, J. M., R. E. Bodmer, and R. A. Mittermeier. 1990. Financial considerations of reserve design in countries with high primate diversity. *Cons. Biol.* 51: 109–114.

Belovsky, G. E. 1988. Extinction models and mammalian persistence. In *Viable Populations for Conservation.* M. E. Soulé, editor. Cambridge University Press, Cambridge.

Bennett, E. L. 1986. *Proboscis Monkeys in Sarawak: Their Ecology, Status, Conservation and Management.* WWF Malaysia, Kuala Lumpur, and New York Zoological Society, New York.

———. 1988. Proboscis monkeys and their swamp forests in Sarawak. *Oryx* 22: 49–57.

———. 1989. *Conservation and Management of Wetland Areas in Sarawak.* New York Zoological Society, New York, and WWF Malaysia, Kuala Lumpur.

Bennett, E. L., and A. C. Sebastian. 1988. Social organisation and ecology of proboscis monkeys (*Nasalis larvatus*) in mixed coastal forest in Sarawak. *Int. J. Primatol.* 9: 233–255.

Bennett, E. L., and M. Walsh. 1988. A wildlife survey of the proposed Matang National Park, Sarawak. Sarawak Forest Department, Kuching. Report.

Caldecott, J. O. 1988. *Hunting and Wildlife Management in Sarawak.* IUCN, Gland, Switzerland, and Cambridge, England.

Choon, C. C., and J. D. Mamit, compilers. 1990. *Briefing Notes and Document for International Tropical Timber Organisation Mission to Sarawak.* Botanical Research Centre, Kuching, Sarawak.

Chung, K. S. 1987. *Estimating the Demand Curves of Bako, Niah and Lambir Hills National Park of Sarawak.* Sarawak Forest Department, Kuching.

Frankel, O. H., and M. E. Soulé. 1981. *Conservation and Evolution.* Cambridge University Press, Cambridge.

Good, L. 1991. *A Management Plan, Niah National Park.* National Parks and Wildlife Office, Sarawak Forest Department, Kuching.

Goodman, D. 1987. The demography of chance extinction. In *Viable Populations for Conservation.* M. E. Soulé, editor. Cambridge University Press, Cambridge.

Gumal, M., and S. S. Tan. 1992. Some aspects of social forestry in Sarawak. Paper presented at the Eleventh Malaysian Forestry Conference, Kota Kinabalu, Sabah.

Kavanagh, M. 1985. Planning considerations for a system of national parks and wildlife sanctuaries in Sarawak. *Sarawak Gaz.* III: 15–29.

Kavanagh, M., A. A. Rahim, and C. J. Hails. 1989. *Rainforest Conservation in Sarawak: An International Policy for WWF.* WWF-Malaysia, Kuala Lumpur.

Kemp, A. C., and M. I. Kemp. 1975. Report on a study of hornbills in Sarawak. WWF-Malaysia, Kuala Lumpur.

Lamb, A. 1991. Orchids of Sabah and Sarawak. In *The State of Nature Conservation in Malaysia.* R. Kiew, editor. Malayan Nature Society/International Development and Research Centre of Canada, Kuala Lumpur.

MacKinnon, J. R., and K. S. MacKinnon. 1986. *Review of the Protected Areas System in the Indomalayan Realm.* UNEP/IUCN, Gland, Switzerland, and Cambridge, England.

McNeely, J. A. 1988. *Economics and Biological Diversity: Developing and Using Economic Incentives to Conserve Biological Resources.* IUCN, Gland, Switzerland.

McNeely, J. A., K. R. Miller, W. V. Reid, R. A. Mittermeier, and T. B. Werner. 1990. *Conserving the World's Biological Diversity*. IUCN, Gland, Switzerland; WRI, CI, WWF-US, and the World Bank, Washington, D.C.

National Parks (Amendment) Ordinance. 1990. The State Attorney General's Chambers, Kuching, Sarawak.

Ngui, S. K. 1991. National parks and wildlife sanctuaries in Sarawak. In *The State of Nature Conservation in Malaysia*. R. Kiew, editor. MNS, United Selangor Press, Kuala Lumpur.

Ngui, S. K., and P. K. Chai. 1985. Development of national parks and nature reserves in Sarawak. Paper presented at the 1985 Annual Forest Department Conference, Sarawak Forest Department, Kuching.

NPWO. N.d. *A Proposal to Establish Santubong National Park*. National Parks and Wildlife Office, Sarawak Forest Department, Kuching.

———. 1987. *A Proposal to Constitute Hose Mountain National Park in the Seventh Division, Sarawak*. National Parks and Wildlife Office, Sarawak Forest Department, Kuching.

———. 1992a. *A Revised Proposal to Protect the Pulong Tau National Park in the Miri and Limbang Divisions of Sarawak*. National Parks and Wildlife Office, Sarawak Forest Department, Kuching.

———. 1992b. *A Proposal to Constitute the Talang-Satang National Park*. National Parks and Wildlife Office, Sarawak Forest Department, Kuching.

Pearce, K. G. 1989. Conservation status of palms in Sarawak. *Malay. Nat.* 43: 20–36

Primack, R. B. 1991. Logging, conservation and native rights in Sarawak forests. *Cons. Biol.* 5:126–130.

Rajanathan, R. 1992. Differential habitat use by primates in Samunsam Wildlife Sanctuary, Sarawak. M.S. thesis, University of Florida, Gainesville.

Sarawak Forest Department. 1991. *Forestry in Sarawak, Malaysia*. Sarawak Forest Department, Kuching.

Schonewald-Cox, C. M. 1983. Guidelines to management: A beginning attempt. In *Genetics and Conservation*. C. M. Schonewald-Cox, S. M. Chambers, B. MacBryde, and L. Thomas, editors. Benjamin Cummings, Menlo Park, Calif.

Soulé, M. E. 1988. Where do we go from here? In *Viable Populations for Conservation*. M. E. Soulé, editor. Cambridge University Press, Cambridge.

Soulé, M. E., and D. Simberloff. 1986. What do genetics and ecology tell us about the design of nature reserves? *Biol. Cons.* 35: 19–40.

UNEP (United Nations Environment Programme). 1992. *Convention on Biological Diversity*.

Watson, H. 1985. *Lambir Hills National Park Resource Inventory with Management Recommendations*. National Parks and Wildlife Office, Sarawak Forest Department, Kuching.

Zainuddin, in prep. *Wildlife Conservation in Hill Dipterocarp Forest in Sarawak*.

13   THE SIGNIFICANCE OF THE TIMBER INDUSTRY IN

THE ECONOMIC AND SOCIAL DEVELOPMENT OF

SARAWAK

Hamid Bugo

The government of the Malaysian state of Sarawak is strongly committed to conserving its forest resources and improving the well-being of all the people of Sarawak. Forest resources support many rural people, and forest maintenance protects the source of income and employment of much of Sarawak's population. The state government has long adopted a forest management policy that regards timber as a strategic resource to be conserved and managed efficiently for the benefit of both present and future generations.

This chapter discusses the timber industry in Sarawak from the government's perspective, considering timber as a strategic resource that is essential to the state economy. The state government recognizes that sustained economic development cannot occur without due consideration to the delicate ecosystem of the forest and the sensitive issue of native land rights, but the forest resources are essential to the economic growth of Sarawak.

SARAWAK'S FORESTRY POLICIES

Sarawak is the largest state in the Federation of Malaysia, with a land area of 124,449 km². About 70% of Sarawak's area is covered with forest. Another 25% is under shifting cultivation; most of this area is secondary forest used by the local people to cultivate hill paddies (rice). The rest is used for permanent agriculture and human settlement (table 13.1). More than half of the forested land is subject to legal strictures limiting the type and intensity of timber harvesting.

Legal protection and sustainable use policies have been in effect in Sarawak for almost forty years; to manage its forest resources effectively and sustainably, the government of Sarawak has identified the extent of these resources, determined their distribution and function, and developed clear policies and management guidelines. Forest resources are categorized into three types: Totally Protected Areas (TPAs), the Permanent Forest Estate (PFE), and

Table 13.1 Major types of land use in Sarawak, 1985 and 1990

Land use classes	1985 (%)	1990 (%)
Forest land (including national parks and wildlife sanctuaries)	70.7	69.9
Shifting cultivation	25.4	25.7
Settled agriculture	3.4	3.8
Other	0.5	0.6

Source: Data from the Sarawak Forest Department.

Note: Percentages are based on a total land area of 12.3 million ha.

State Land Forests (SLFs) (table 13.2). The TPAs include national parks, wildlife sanctuaries, and wildlife rehabilitation centers created under the National Parks Ordinance (Sarawak Chap. 127, 1956) and the Wildlife Protection Ordinance (Sarawak Chap. 128, 1958), respectively. These areas, which encompass 290,000 ha (3.3% of the forested land in Sarawak), are designated for biodiversity conservation (Abang Morshidi and Gumal, this volume; Kavanagh, Rahim, and Hails 1989; Primack 1991a). Steps are being taken to increase the land area of the TPAs to 1 million ha. The PFE, mandated by the Forest Ordinance (Sarawak Cap. 126, 1954), is designated for sustainable forest management. The PFE covers 4.5 million ha, or 54% of forested land, and will be increased to 6 million ha (72%). The remaining 1.7 million ha are designated as SLFs and are earmarked for conversion to agriculture and other purposes.

Timber-Harvesting Policies and Regulations

In line with the Sarawak Forest Policy, which was formulated and adopted in 1954, the objectives of forest management in the PFE are to optimize the use of the forest resources, to regulate timber harvest on a sustained-yield basis, to remove timber in an orderly manner with the minimum damage to the residual stand and the environment, and to regenerate and rehabilitate harvested forests to improve the forest stocking of valuable timber species by proper silvicultural techniques (ITTO 1990; Primack and Hall 1992).

To achieve these management objectives, each forest concession, known as a Forest Management Unit (FMU), has a Forest Management Plan. The Forest Management Plan contains the following items:

1. Description of the physical features of the FMU
2. Objectives of management

Table 13.2 Natural forest cover in Sarawak, 1985 and 1990

Type of Area	1985 (ha)	1990 (ha)
Natural forest cover	8,455,314	8,382,924
Designated as permanent forest estate	4,152,157	4,597,366
Designated as national parks	76,949	113,955
Designated as wildlife sanctuaries	174,851	174,851

Source: Data from the Sarawak Forest Department

3. Prescriptions for how the FMU is to be harvested
4. Species to be removed and their diameter limits
5. Annual allowable harvest
6. Penalties for harvesting damages
7. Logging road specifications
8. Responsibilities and duties of the concession holder
9. Responsibilities and duties of the branches of the Sarawak Forest Department

The Forest Management Plan is part of the Forest Timber License and is a legal document. The intent of this plan is to state clearly the conditions of the license, to present guidelines for harvesting, and to promote sustainable forest use.

Harvesting of commercial timber in Sarawak is concentrated in the hill mixed dipterocarp forest and peatswamp forest. A selective felling system that sets limits on the diameter of trees to be harvested has been adopted in the PFE to ensure sustained yield and efficient harvesting operations. The objective is to remove all the mature timber trees in a single felling operation, leaving behind a residual stand with a sufficient number of healthy, fast-growing trees in the intermediate diameter classes to form the next crop. Under this system, different diameter limits are imposed for each forest type. In the hill mixed dipterocarp forest, the diameter limits are 60 cm for dipterocarp species and 45 cm for other timber species. In the peatswamp forest, the diameter limits are 40 cm for ramin trees (*Gonystylus* spp.) and 50 cm for other species. Cutting cycles of 25 years are in effect for the hill mixed dipterocarp forest; the cycle is 45 years for the peatswamp forest. The adoption of cutting cycles ensures sufficient time for a harvested forest to regenerate.

Conservation in Harvested Forests

The forest policy of Sarawak is intended to create permanently reserved forestland sufficient for environmental protection and to supply in perpetuity all forms of forest produce. In addition, forestry practices are designed to manage production forests to obtain the highest possible revenue compatible with the principle of sustained yield. Ecologically and environmentally sound forest management practices have been developed to ensure forest renewal and sustained yield. The Sarawak Forest Department emphasizes the need for forest conservation and environmental protection in the harvested forests. For this reason, forest harvesting in the PFE is prohibited on slopes greater than 60% to prevent severe soil erosion, and logging operations must submit plans demonstrating a good-faith effort to minimize damage to watersheds in forests where logging is permitted.

CONFLICTS AND CRITICISMS

Given its forest management policy, Sarawak will probably be able to maintain a forest cover on about 70% of its land area. Yet for a number of reasons Sarawak (Malaysia) has been subjected to intense criticism by environmental non-government organizations (NGOs) (INSAN 1992). First, some NGOs assume, because Sarawak does not practice replanting, that the felled trees are not replaced. Through selective felling, however, younger trees are allowed to grow up naturally and replace the harvested mature trees. In fact, opening of the canopy through extraction of larger trees may actually encourage growth of young trees and saplings, which would produce a more naturally diverse forest than one composed of planted species.

Second, political disagreements exist in Sarawak, as in any country. Factions occasionally use the international media to advance their own agendas. For example, many people outside Sarawak believe that the government has deprived the nomadic Penans of their home and their livelihood as the forests have been harvested. To some people outside Malaysia, the Penan tribe, one of many in Sarawak, represents a classic case of a small-scale egalitarian society ruthlessly oppressed by the forces of modern government (Hansen 1989; INSAN 1992). Most Penans, however, have voluntarily adopted sedentary agriculture. Only about 400 out of 10,000 Penans are nomadic and live in the forest (fig. 13.1). Though the nomads are clearly a minority, the government has taken their needs into consideration and has set aside certain areas as biosphere reserves for their use (Primack 1991a, 1991b). Despite this action, however, a small, vocal faction of disgruntled Sarawakians, claiming to represent the natives, use the environmental NGOs to proclaim that the rights of native people are being ignored or circumvented by the government. Many people outside Sarawak take these claims at face value, not realizing that there are often ulterior motives, personal and political, for such accusations. Government policy regarding the nomadic tribes has always been to give them the choice between their traditional way of

FIGURE 13.1 Nomadic Penans use specialized forest products to maintain their way of life. In this photograph a Penan man carrying a blowgun stands in front of an ipoh tree (*Antiaris toxicaria*). Poison sap, collected by making diagonal cuts in the tree trunk, is used to coat blowgun darts, with deadly effect in hunting. Photograph by R. Primack.

life and life in modern settlements; further, the government provides facilities for both options and assists in the transition period if tribespeople choose to settle.

Finally, Sarawak shares a dilemma with many other young nations in the modern world: how does a state develop its economy and improve its standard of living without risking the health of its natural resources? Many outside the state would have the government set aside all natural forests in perpetuity, but this solution is unrealistic and burdensome for the people of Sarawak, who need jobs and income.

Added to this problem is the consideration of the rights of indigenous people. Throughout the world, the issue of native rights is a volatile subject, and controversies naturally attract attention from the international media. Many false perceptions may develop from a lack of understanding of the complicated and changing cultural landscape in Malaysia. Well-meaning individuals from outside Sarawak have created sensational stories that do not necessarily reflect the real situation in the country. The issues are very complex, and a casual visit to indigenous people's villages or camps is not sufficient to create a true understanding of the problems the villagers face. For example, the absence of a school in a remote village should not automatically be equated with government neglect of the people of that village. Instead, perhaps no teachers are willing to relocate to a remote area, or perhaps there is no safe, suitable site on which to build, or perhaps local people prefer traditional methods of education and do not want a government-operated school in their community.

CONTRIBUTIONS OF TIMBER INDUSTRIES TO SARAWAK'S ECONOMY AND SOCIETY

Population and History

According to a census conducted in 1991, Sarawak has a population of 1,648,217, with a growth rate of 2.6% annually in the 1980s. Population density is low, averaging of 13 persons/km²; however, the population is scattered unevenly, with a relatively heavy concentration in the coastal zone that quickly thins out toward the mountainous interior areas.

The population, though small, is made up of numerous ethnic groups. The main ones are the Iban (29.5% of the population), Chinese (29.3%), Maylay (20.8%), Bidayuh (8.4%), and Melanau (5.5%). The "other" category of indigenous people (5.5% of the population) is composed of no fewer than twenty minor indigenous groups, each with its own customs and language or dialect. Apart from the Chinese, all ethnic groups are natives of Sarawak. The Chinese are found mostly in the urban centers, and the Malay and Melanau mainly occupy the coastal area and fringes of the towns. The other native groups traditionally occupy the hilly and mountainous regions of the state; with the passage of time and increased modernization, a sizable proportion of these groups have migrated downriver and into the towns. Most, however, prefer to remain in their traditional territories on the hill and mountain slopes and at the fringes of secondary forests. Here they practice traditional shifting cultivation in hill paddies, combined with small plots of cash crops and other food crops.

Low-productivity subsistence farming is prevalent. Many subsistence farmers, especially those with minimal access to towns, must depend on forest resources, such as wild animals, fish, and vegetables, for a portion of their food. As more of these rural settlements become accessible through the construction of roads and logging tracks, however, the degree of their inhabitants' dependence on the forest as a source of food and sale items decreases. Nearly all the Bidayuh of Bau and Serian Districts, for instance, have become cash croppers and wage earners in the 1980s and 1990s.

The Penan are an exception to this trend. Numbering approximately 9,745, the Penan are among the smallest of the native groups, and they are the only group with a majority still relying on the forest for subsistence. Only about 400 individuals (4% of the population), however, are truly hunters and gatherers who roam the old-growth forest of the Tutoh and Limbang watersheds in small groups of about 30 persons (see fig. 13.2 for their locations). These people are the truly nomadic Penan. About 25% of the remaining Penan population have adopted a sedentary mode of life. Members of this subgroup live like the other settled native groups: they grow paddy and other cash crops for the market, and some earn wages. The forest is no longer an important source of food for this segment of the Penan. The rest of the Penan population (about 70%) is in a transitional stage; their way of life combines the customs of the nomads and the settled people. They grow basic food crops, such as hill paddy, tapioca, and some vegetables, using simple farming methods. The rice supply of most families lasts only a few months. Partly for this reason, Penans still depend on the forest not only for food but also for products, such as rattan, used for barter and handicrafts.

The British colonial administration encouraged the Penan to settle downriver, closer to administrative centers, to facilitate the provision of basic services. The process of resettlement, which began more than half a century ago, is ongoing, with most of the Penan population still in the transitional stage. This stage is always difficult; members of the group have to acquire new skills and patterns of behavior and must adjust to a new environment (INSAN 1992; ITTO 1990; Primack 1991a). Any disturbance or disequilibrium in the environment is bound to have grave emotional and psychological effects on the community members. For this reason, the state government has carefully avoided forcing the Penan to join the settled way of life common to the rest of Sarawak. At the same time, the government has provided the nomadic Penan with adequate facilities to assist them in the transition to settled life if they choose.

Economic Contributions of the Forestry Sector

The economy of Sarawak in 1990 was led by three primary industries, which constitute 55% of the state Gross Domestic Product (GDP): mining and quarrying, forestry, and agriculture (table 13.3). The forestry subsector was the second largest sector of the economy in 1990, after mining and quarrying. In fact, forestry has always been an important sector, second to agriculture in the 1960s and early 1970s, and surpassed only by mining and quarrying, mainly of oil and natural gas, from the mid–1970s to 1990. The state government does not intend to maintain the structure of the economy in its present form, however; a structural change will make forestry the leading sector of the economy.

From 1980 to 1990, Sarawak's economy grew about 6.6% annually, despite the worldwide recession. In real terms, the economy grew from $MR3.5 billion in 1980 to $MR6.7 billion in 1990 (1 Malaysian ringgit equals 0.37 U.S. dollars). Most of this growth was in the mining and quarrying, forestry, and

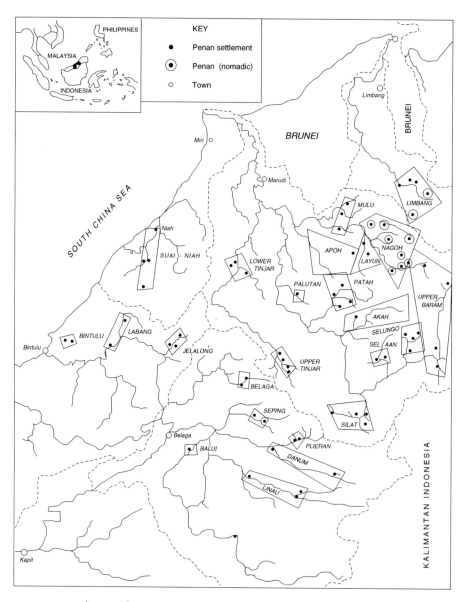

FIGURE 13.2 Areas in Sarawak occupied by nomadic and semi-nomadic Penan groups. Dots show the center of each territory, and boxes enclose the areas that the Penans use. Territories are centered around rivers and streams.

Table 13.3 Sarawak Gross Domestic Product by industry, 1981 and 1990

Industry	1981 (% share)	1990 (% share)
Agriculture, livestock, and fishery	12.8	9.3
Forestry and logging	12.9	14.8
Mining and quarrying [a]	29.6	32.8
Manufacturing	6.5	13.0
Construction	5.9	3.3
Wholesale and retail	8.4	7.5
Transport, storage, and communication	3.8	4.8
Social, personal, and other services	20.1	14.5

Note: Percentages are based on GDPs of 3,679 million Malaysian ringgit in 1981 and 6,667 million Malaysian ringgit in 1990 ($MR1 = US $0.37).
[a]Mainly oil and natural gas.

manufacturing sectors. During this period the forestry and manufacturing sub-sectors expanded by 7.6% and 13.6% per annum, respectively. Though the forestry subsector contributed only 14.8% to the GDP in 1990, its significance to the state's socioeconomic stability surpassed this figure. Its real contribution is best seen in terms of state revenue and employment. At the present rate of royalty imposed on log production (which is relatively low compared to those imposed by neighboring states), royalties from logs contribute about 50% of state revenue. In 1990, for instance, royalties and other duties from logging were $MR705.1 million, compared to $MR310.9 million derived from petroleum and gas royalties.

Roughly 85% of the estimated 18.8 million m³ of logs produced in Sarawak in 1990 were exported in raw form, primarily to East Asian countries such as Japan, Taiwan, Hong Kong, and South Korea. The total value of exported timber was $MR2.9 billion, constituting 25% of export income for that year (table 13.4). Although the export of logs benefits Sarawak and Malaysia in terms of foreign exchange earnings, the government does not intend to perpetuate this trend. Timber is an important resource; together with petroleum and gas, it will form the backbone of the industrialization program intended to bring about structural change in the economy. The state aims to increase the manufacturing sector's contribution to the state GDP to about 20% by the year

Table 13.4 Value of raw timber exports from Sarawak in comparison with total exports for 1981, 1985, and 1990

Year	Total export (million $MR)	Export of raw logs (million $MR)	Share (%)
1981	4,517	812	18
1985	8,447	1,404	17
1990	11,284	2,883	26

Note: $1MR = US $0.37.

2000, a 5% increase from the 1992 level. To achieve this objective, the government implemented an export quota on logs in 1989; by 2000, at least 50% of the logs produced will be retained for domestic processing, to take advantage of value-added employment and other economic linkages.

Since this quota policy was announced, the response from the private sector, both local and foreign, has been overwhelming. The total amount of investment in the sector has increased from only slightly more than $MR4 million in 1981 to more than $MR100 million in 1989, a year after the policy was unveiled. In the 1970s and early 1980s, most investment was in sawmills and small furniture factories that were mainly for domestic consumption (fig. 13.3). Toward the late 1980s, however, larger factories for the production of plywood and veneer became more prevalent. The production of plywood and veneer has increased significantly, rising from 19,519 m³ and 3,157 m³, respectively, in 1981 to 341,800 m³ and 163,900 m³, respectively, in 1991. During the period 1989–91, 82 orders for the production of plywood and veneer were received. This number was higher than the number of all the orders received in the earlier years of the 1980s combined.

Although the trend in industrialization is encouraging, the government nevertheless is working toward further downstream processing to produce higher value-added products for export, such as furniture and furniture parts, laminated board, and prefabricated housing materials. A number of industrial estates have been set up, mainly for timber processing; this policy and the policy of export quotas on logs are aimed to industrialize the economy.

Employment

The timber industry provides attractive employment to Sarawak's growing population. In 1990, logging employed around 60,000 people, including workers involved in surveying and preparing the areas for logging, felling trees, debarking, and moving the logs to the log ponds. Wages range from

FIGURE 13.3   Large sawmills in Sarawak produce timber products for export and domestic consumption. Photograph by R. Primack.

around a thousand ringgit per month for unskilled workers to a few thousand ringgit for experienced tractor drivers. As the logging operations are invariably located in the deep forest, they are generally not attractive to urban youths; thus most workers in logging camps are rural people. The timber-based manufacturing industries (sawmills and plywood factories) are another important source of employment. At the end of 1990, at least 23,000 people were employed in timber-based manufacturing throughout the state. More than half of these jobs are in the production of sawed timber, but other products, such as plywood, veneer, wooden molding, laminated board, and other industries, contribute nearly 9,000 jobs.

Numerous employment opportunities are available in related activities, such as trucking and shipping. At the same time, secondary employment in service industries supported by the buoyant forestry sector is an option. The increased demand for public transport service, especially water transport, and retail trades help make Sarawak's rural economy a vibrant one.

Social Impacts of the Forestry Sector

One of the most significant problems facing Sarawak's society is its high poverty level. In 1976, over half the population was considered impoverished by government standards, with a high proportion of poor families (64.4%)

in rural areas. The steady decline since then in overall poverty, particularly in rural areas, is directly related to growth in the timber industry. Workers in timber camps and factories are mainly native people from rural areas. Cash remittances from these workers constitute a significant portion of rural household incomes. Because the prices of most internationally traded agricultural commodities, such as rubber, pepper, and cocoa, have been depressed since the late 1980s, timber and timber-related industries supply the greatest possible income for rural families. In 1990, for instance, the average income of rural households was $MR1,045 per month, as compared to $MR843 per household in 1984. Because of this increase in household income, poverty in rural areas declined from 37.2% in 1985 to 24.7% in 1990. Statewide, poverty declined from 31% to 21% during the same period. Economic projections to 1995 indicate that the steady decline in poverty will continue, reaching an estimated low of 16%.

Other social indicators also suggest that living conditions for rural people improved significantly in the 1980s. The proportion of the rural households that have indoor plumbing and electricity increased dramatically in a single decade: in 1980, 20% of rural households had running water and 30% had electricity; by 1990, 47.3% had water and 46% had electricity. Similarly, basic educational and medical services are becoming more readily available to the rural population, including those in isolated settlements, through boarding facilities and flying doctor services. Two service centers have already been completed for Penan communities—one at Long Kevok in the interior of Baram District and another at Lusong Laku in Belaga District. Each center consists of a primary school, a clinic, and an agriculture station, and both are staffed by full-time resident officials. Each center serves a population of 200–400 Penan households in several nearby settlements. In this way, essential services are made accessible to Penan communities. Another four centers serving the needs of these communities and their neighbors are planned.

Because of their employment in timber-related industries, many native people have been technically trained. Skilled workers, such as truck and tractor drivers, machine operators, and road construction workers, were not previously available in rural areas; these skills are also in high demand in other industrial sectors of the state economy. Logging operations provide native people with invaluable training and experience in commerce and trades.

In many remote settlements, a logging operation means not only availability of badly needed jobs but also improved accessibility to formerly isolated settlements. Throughout the upper Baram and Kapit Districts, where logging activities are prevalent but access is difficult, timber tracks crisscross the landscapes. Although these tracks are primarily transporting logs from operation sites to log ponds, they also transport people and goods from their isolated settlements to the nearest administrative and commercial centers. Realizing their importance, the state government is considering the feasibility of upgrading some of these tracks into development roads.

PROBLEMS IN ECONOMIC DEVELOPMENT

A number of problems hinder Sarawak's efforts to develop a sound state economy, among them a sparse population scattered over wide areas of difficult terrain, a poor transportation system, environmental concerns, and a complex land tenure system. In this chapter, however, I deal only with the last two issues, as they relate directly to logging activities in the state.

Logging of the tropical rainforest undeniably disturbs the surrounding areas, as well as the local people inhabiting the forest fringe. Logging activities often create conflicts between the loggers and the native people over such issues as the natives' rights to the land and forest, concern for sources of food and water, and destruction of their sacred ground, especially burial grounds (ITTO 1990; Kavanagh, Rahim, and Hails 1989; Primack 1991a) (fig. 13.4). Although it is not possible to harvest the tropical rainforest without disturbing the environment, Sarawak's laws require that the Forest Department and log-

FIGURE 13.4 A Penan man is arrested at a blockade of a logging road, part of an effort to halt logging, in an area claimed by the nomadic Penan. Photograph by the Sarawak Forest Department.

gers take special care to minimize the effects of logging on the surrounding areas. The Forest Department requires that a proper forest engineering plan be drawn up and implemented to ensure that logging roads are constructed properly to minimize soil erosion and siltation of rivers and streams. Similarly, water catchment areas and sacred grounds must be clearly marked on the plan. These plans can help ensure that important sites will not be disturbed during logging.

Despite the mechanisms in place to minimize damage to the landscape and to prevent conflicts between loggers and local people, problems sometimes arise. Local people accustomed to free use of state-owned land often feel that they own the land. Logging operations, however, occur only in primary forest that comes under the Reserve Land and Interior Area Land classification of the Land Code. Both categories of land are state-owned; if native people legitimately claim customary rights, then by law the claimed area is automatically excluded from the licensed area, unless the native people agree that the area need not be excluded. In such cases, the logging companies are required to compensate the claimants in the form of goodwill money. Similar compensation or goodwill money is also paid for damage to fruit trees and burial grounds. The amount of compensation is often a further cause of dispute between the two parties.

Native customary claims become an issue or a constraint to development when the boundaries of the claims are not clearly marked. Damage to burial grounds, for instance, frequently occurs because these areas are difficult to identify. In addition, certain segments of the native population do not accept the Land Code, which does not recognize customary rights on land first used by native people after January 1958. Native people are still making claims to lands, especially forestlands, that the government has identified for logging or commercial agriculture. Such claims, valid or otherwise, are bound to create conflicts between the loggers and the claimants (Primack 1991a).

STRATEGIES FOR ECONOMIC SECURITY

Since Sarawak achieved independence in 1963, the government has embarked on a conscious effort to develop the state and its people. In line with the National Development Policy and the Second Operational Perspective Plan, the goal for the 1990s is to achieve a balanced development for sustained economic growth and equitable distribution of wealth. The economy must be restructured to reduce dependence on the export of primary commodities. The government intends for at least 20% of the state GDP to come from the manufacturing sector by the turn of the century. The state also aims to improve physical and social infrastructures and public utilities, both to spur development in areas with economic potential and to supply rural areas where such facilities are not readily available. One goal is to provide at least 75% of rural households with electricity and pipe water by the end of the 1990s.

A large proportion of development is planned for rural areas, in order to distribute wealth more equitably. Under the development plan, 47% of the funds are allocated for projects in rural areas, and another 35% are for projects to benefit both rural and urban settings. These funds are mainly for infrastructure, such as clinics, schools, public utilities, and development of smallholder agriculture.

Sarawak is rich in natural resources, including timber, crude oil, gas, and other minerals. In addition, it has vast tracts of land suitable for commercial agriculture that are not yet fully farmed. Most raw materials produced in Sarawak are exported, depriving the state of trade in the more valuable end products and limiting potential employment opportunities. Efforts to restructure the economy will center on three main areas, namely, downstream processing of natural resources (particularly timber, petroleum, and natural gas), tourism, and commercial agriculture. The manufacturing subsector and the service sector will become the mainstays of the economy. In view of the potential of these industries, the state government wishes to promote and facilitate the speedy and orderly development of the manufacturing industries based on the state's available resources, involving both local and foreign capital and expertise.

In line with this policy, the state has embarked on a comprehensive industrialization program by establishing a separate ministry in 1987 to implement the program. Sarawak shares the national vision that Malaysia will be a developed country by 2020, and thus Sarawak will be a developed state. By 2020, the secondary and tertiary sectors will be the dominant features of the state economy, and the people will enjoy a higher standard of living with easy access to modern services and facilities.

The Role of Timber Industries in Sarawak's Future Development

Timber is a strategic resource for Sarawak's industrialization program; its importance supersedes that of crude oil and gas. Timber-related industries are most likely to provide a jump start for the government's ambitious development program. Development cannot take place without capital for building roads, factories, and communications infrastructure; the state government views timber as the best source of the income needed for initial startup costs. Though petrochemical products account for more than 50% of the value-added portion of the manufacturing sector's contribution to the state GDP, petrochemical industries are capital intensive and do not generate sufficient job opportunities to meet the needs of the growing population. Timber has already provided the state government with a major portion of the revenue needed to finance its development program; revenue from logging accounts for about 47% of the state revenue since 1988 (table 13.5). A drastic reduction in timber income would adversely affect the state's ability to operate and implement programs for the people's benefit.

Nevertheless, the state government does not intend to sustain log production at the level of 1991. Rather, it plans to maintain its income by reorganizing the industry, limiting log production to a level that will ensure sustainable development (ITTO 1990). Unlike petrochemical industries, timber-based industries can take many different forms and sizes, in terms of both capital investment and sophistication. They can be, for example, a simple, family-run business producing furniture for the local market. This form of undertaking does not require a high capital investment or highly skilled workers. Until mid–1991, more than 90% of all the timber-based establishments in the state were of this type—small or medium-sized factories operated by individuals with the help of family members and a few hired workers. They produce furniture (mainly for the local market), sawed timber, moldings, and dowels. This enterprise provides ample opportunities for local people to be involved directly in the industry.

Ongoing Changes in Sarawak's Economy

Large timber-based establishments are a recent phenomenon developed in response to government policies designed to reduce log export and increase local processing. These giant enterprises mainly produce plywood, veneer, and laminated boards for export. One of these plywood factories can employ more than a thousand workers of various skill levels. Most are owned by local logging operators who are diversifying their business ventures into more value-added products. The recent increase in log production is mainly due to clear-felling for the purpose of cultivating oil palm, which indicates that the proceeds from logging are being used locally to help the state develop and generate more jobs.

Sarawak's forests, apart from being a source of timber and other products, are attractive to tourists. The diverse flora and fauna, as well as waterfalls, limestone caves, and beaches, are invaluable resources. The government is marketing the state as a tourist destination combining adventure, ecology, and folk culture. A tourism plan is being prepared that will identify potential tourist

Table 13.5 State revenue generated from forest resources, 1988–1991

Source	1988 ($MR)	1989 ($MR)	1990 ($MR)	1991 ($MR)
Royalties, permits, premium, and assessment	325,100,268	629,337,230	703,495,333	666,234,777
Export duty on timber	186,517,737	22,846,547	21,427	82
Export duty on forest produce	499,622	739,922	1,562,390	733,991
Total from forest	**512,117,6237**	**652,923,699**	**705,079,150**	**666,968,850**
Total state revenue	**1,093,407,371**	**1,251,579,385**	**1,467,033,577**	**1,660,259,830**
% Contribution of Forestry Sector	**46.8%**	**52.2%**	**48.1%**	**40.2%**

Note: $1MR = US $0.37.

attractions and recommend appropriate strategies to develop and manage state resources on a sustainable basis.

The response to the industrialization program so far has been encouraging. In the timber-based industries alone, the number of approved projects has increased rapidly in the 1980s. The paid-up capital for the wood-based manufacturing industry has increased from $MR4.2 million in 1981 to $MR3.1 billion in 1990. Potential ventures in this industry include quality furniture, laminated board, specialized molding and joineries, and prefabricated housing materials. A similar trend is occurring in the petrochemical industries; multibillion-ringgit natural-gas processing plants are being constructed in Bintulu. Investors have also shown interest in other petrochemical projects, including polypropylene, copralactam, ammonia, melamine, and compound fertilizer. Other resources for manufacturing activities are silica sand and Kaolinic clay, important raw materials for glass and porcelain manufacture that are mined for export only (in 1988, Sarawak produced 391,000 tons of silica sand, all for export). The creation of the first free trade zone in the state offers opportunities for such non-resource-based industries as electronics and other high-tech trades.

The potential of agricultural industries has not yet been fully explored. To further support the development of this sector, the government is actively promoting commercial agriculture on large plantations and mini-estates. At least 10,000 ha of land will be developed and planted annually with commercial crops. To achieve this objective, the government will continue to provide and improve basic infrastructure, especially roads, and other services in order to encourage the private sector to invest in plantation agriculture. Oil palm, sago, pineapples, and certain tropical fruits grow well in the state, and the products from these crops can form the basis for the processing activities. The potential of other crops is being investigated.

CONCLUSION

The rural population of Sarawak has responded favorably to the development policy and programs of the government. A group of lecturers at the Universiti Pertanian Malaysia polled members of all ethnic groups from both urban and rural areas. Their study indicates that the people of Sarawak have a positive view of government programs and projects (UPM 1991). For instance, 95% of the respondents indicated that the agriculture projects are useful and must be continued, and another 80% indicated that they want the government to develop their native customary lands. This finding is revealing—the native people, who consider their lands an important hereditary property from which they would not want to depart, are convinced that this property must be developed for their benefit. Thus land development agencies, such as SALCRA and FELCRA, find themselves unable to cope with the many requests of the native people to have their lands developed for commercial agriculture.

The Penan, the least sedentary group, also want development. They desire a way of life similar to that of their more settled neighbors, mainly the Kayan, Kenyah, and Berawan. Many Penans wish to own modern household items and have easy access to modern facilities and services, especially schools and clinics. Government officials who visit Penan settlements receive requests for a variety of modern items, such as electric generators, outboard engines, rice mills, building materials, schools, and clinics. In a working paper presented at a Penan development workshop in Marudi, Peter Brosius (1992), an anthropologist with extensive experience working with the Penan, pointed out that most Penans in Sarawak want to live settled lives, just like other Malaysians. But he emphasized that the community must change more slowly than progressive and settled communities and that the pace of development must be adjusted to suit the community's capacity to deal with change.

Sarawak faces an extremely complicated set of circumstances. In an ideal world, the government might find a way to reconcile conflicting needs: the needs of the population for income and the desire to maintain intact the old-growth forests. Given its abundant resources, Sarawak's economic prospects are bright, but some of the development must come at the immediate expense of a portion of the state's natural resources. The government of Sarawak is fully aware of lessons learned so painfully by the West in the nineteenth century and more recently by other tropical countries: the tropical forest is an invaluable asset, both in the economic and socioaesthetic sense, and to mine its resources for short-term gain at the expense of future prosperity would serve little purpose. State officials know that they are duty-bound to manage the forest properly for the long-term development of the state. Sarawak's future economic security depends on careful use of its renewable resources. In other tropical countries, such as Brazil, poverty and population pressure have lead desperate people to wantonly destroy valuable forestland by converting it to farmland. Sadly, the land becomes infertile after only a few years, and the farmers then destroy more forest. The government of Sarawak is convinced that by conducting the timber industry with care and forethought such extremes can be prevented (fig. 13.5). Well-managed forests combined with continuing industrial development will ensure not only the prosperity of the people of Sarawak but the continued health of the forests.

FIGURE 13.5   In the background is a well-managed forest near Miri, Sarawak, which has been selectively logged three times in fifty-five years and still maintains good forest cover and the potential for future wood production. In the foreground, farmers have cut down and burned the trees to practice shifting cultivation. Photograph by R. Primack.

REFERENCES

Abang Morshidi, A. H. K., and M. T. Gumal. This volume. The role of totally protected areas in preserving biological diversity in Sarawak.

Brosius, J. P. 1992. Perspective on Penan development in Sarawak. *Sarawak Gaz.* 119, 1519.

Cramb, R. A., and R. H. W. Reece, 1988. Development in Sarawak. Center of Southeast Asian Studies, Monash University.

Hansen, E. 1989. *Stranger in the Forest: On Foot across Borneo.* Viking Penguin, New York.

INSAN (Institute of Social Analysis). 1992. *Logging against the Natives of Sarawak.* INSAN, Petaling Jaya, Malaysia.

ITTO (International Tropical Timber Organization). 1990. The promotion of sustainable forest management: A case study in Sarawak, Malaysia. Report.

Kavanagh, M., A. A. Rahim, and C. J. Hails. 1989. Rainforest conservation in Sarawak: An international policy for WWF. WWF-Malaysia, Kuala Lumpur. Report.

Langub, J. 1989. Some aspects of life of the Penan. *Sarawak Mus. J.* 40, 61, special issue no. 4, part 3.

Mamit, J. D. 1992. Clarifying controversies on forest harvesting in Sarawak. Sarawak Forest Department, Kuching. Memo.

Primack, R. 1991a. Logging, conservation and native rights in Sarawak forests from different viewpoints. *Borneo Res. Bull.* 23: 3–13.

———. 1991b. Logging, conservation and native rights in Sarawak forests. *Conserv. Biol.* 5: 126–130.

Primack, R., and P. Hall. 1992. Biodiversity and forest change in Malaysian Borneo. *BioSci.* 42: 829–837.

UPM (Universiti Pertanian Malaysia). 1991. Kajian persepsi rakyat Sarawak terhadap program pembangunan. UPM, Serdang. Report.

14 TIMBER TRADE, ECONOMICS, AND TROPICAL FOREST MANAGEMENT

Jeffrey R. Vincent

Since the end of World War II, one tropical country after another has experienced boom-and-bust forest sector development (Gillis 1988; Repetto and Gillis 1988; Vincent and Binkley 1992). The boom is marked by rapid logging of old-growth forests (fig. 14.1), which furnishes large volumes of timber for export as logs or, in countries with log export restrictions, as primary processed products (sawnwood and plywood). The bust occurs when the old-growth timber is depleted. Second-growth (logged) forests are generally not managed intensively, and they cannot sustain harvests as high as those in the original old-

FIGURE 14.1   The use of chainsaws, tractors, and other machinery has increased the rate of logging in dipterocarp forests. Photograph by the Sarawak Forest Department.

ьwth forests. Logging and processing industries enjoy great profits during the boom, but the profits inevitably dwindle. The boom typically lasts no more than a decade or two, depending on the size of the country's forest resources.

The boom-and-bust pattern emerged in Ghana and other countries in West Africa in the 1950s and 1960s. It became even more apparent in the 1970s and 1980s after trade shifted toward Southeast Asia and expanded in volume. In Southeast Asia, such countries as Thailand and the Philippines have already gone bust. In virtually all remaining countries the boom is either cresting or waning (Gillis 1988; Scott 1989; Sricharatchanya 1989). This track record seems to indicate that the tropical timber trade is incompatible with sustainable timber production in tropical forests.

EXAMINING THE BOOM-AND-BUST PATTERN

The prevalence of the pattern is often attributed to three factors: (1) developed countries' exploitation of the timber resources in tropical countries, (2) high import barriers by developed countries against processed timber products from tropical countries, and (3) low international prices for tropical timber products. Proponents of this theory reason that consumption in developed countries drives the boom. The development of processing industries in tropical countries is inhibited by import barriers that reduce export earnings and lower the value of the forests. Low prices reflect market manipulation by developed countries, which have supposedly rigged the international trade system to work against the interests of tropical forestry. By reducing the value of tropical forests, for example, trade barriers lower the effectiveness of incentives to manage them for sustained timber production.

The truth of these assertions needs to be examined. The first factor is not supported by statistics on the trade. In 1989 only about a third of the industrial roundwood harvested in tropical countries was exported either as logs or as processed products (table 14.1). Most tropical timber is consumed in the internal markets of the tropical countries. Moreover, much of the exported timber goes to other developing countries (FAO 1991). Because their economies and populations are growing rapidly, developing, not developed, countries account for most of the increase in global consumption of wood products—all products, not just tropical—that occurred in the 1980s or is forecast to occur in the future (Cardellichio et al. 1989; Dykstra and Kallio 1987a; FAO 1988, 1991).

Regarding the second factor, import tariffs against processed wood products certainly exist in developed countries. Like most tariffs, however, they have decreased markedly because of the various rounds of negotiations related to the General Agreement on Tariffs and Trade (GATT) (Olechowski 1987).

Table 14.1 Production and trade volumes for wood products in tropical countries in 1989

	Production	Exports	Imports
Solid-wood products			
(figures in thousands of m³)			
Industrial roundwood[a]	306,256	34,199	12,767
Sawnwood	89,012	13,400	11,691
Wood-based panels[b]	21,200	12,050	3,016
Fiber products (figures in thousands of tons)			
Wood pulp	9,164	1,827	2,607
Paper and paperboard	20,923	2,548	6,904

Source: FAO 1991.
Note: Tropical countries are those classified as "developing" in FAO 1991, excluding China and others whose land areas are predominantly in the temperate zone.
[a] Logs and pulpwood.
[b] Mainly veneer and plywood, but also particle board and fiberboard.

Developed countries' tariffs on wood products are generally comparable to, or lower than, their tariffs on most products, and they are generally lower than corresponding import tariffs in developing countries (table 14.2). Studies indicate that import barriers on wood products in developed countries have only modestly decreased the trade volume for most wood products. Although their removal would increase trade volumes, much of the increase would be captured by exports of softwoods and hardwoods from temperate countries (Dykstra and Kallio 1987b; Olechowski 1987).

The third factor, low prices, is better supported by trade statistics. Since the end of World War II, average export prices for tropical hardwood logs and sawnwood have been substantially lower than corresponding prices for temperate hardwood products (table 14.3). A more innocuous and convincing explanation than market manipulation, however, is that most tropical timber exports compete with commodity products made from temperate hardwoods and softwoods rather than with fine temperate hardwood products. Most tropical timbers have commodity end uses for which there are ample substitutes. Their prices are lower on average than those of temperate hardwoods, but they match or exceed those of temperate softwood products. Thus tropical timber prices are comparable to the prices of the temperate timber products they actually compete against.

This last statement does not mean that prices are an unimportant factor in boom-and-bust forest sector development. In fact, prices—in particular, stumpage values, the value of timber "on the stump"—are crucial, not be-

Table 14.2 Import tariff rates in the early 1980s, expressed as percentages of value

	EC (%)	USA (%)	Japan (%)	LDCs[a]
Wood products				
Wood in the rough	0.0	0.0	0.0	14.4–34.1
Primary wood products	1.9	5.6	7.4	16.2–57.8
Secondary wood products	1.5	1.7	4.8	24.1–73.1
Other products				
Machinery and appliances	4.4	3.2	—	—
Textiles and textile articles	5.6	14.7	—	—
Footwear, headgear, etc.	6.6	12.2	—	—
All items	2.4	2.9	—	—

Sources: Bourke 1988; Olechowski 1987.
Note: Rates for wood products in the European Community (EC), the United States of America (USA), and Japan are for imports from developing countries (LDCs); all other rates are for imports from all sources.
[a] Range of the averages for developing countries in Africa, the Americas, and Asia.

Table 14.3 Export prices for tropical hardwood products relative to those for temperate products: averages for 1945–88

Tropical hardwood product	Corresponding temperate product	
	Hardwood	*Softwood*
Logs (from Asia)	0.58	0.98
Logs (from Africa)	0.88	1.50
Sawnwood	0.73	1.52

Source: Analysis of post–World War II data from various issues of the *FAO Yearbook of Forest Products*.
Note: Averages are price of tropical product divided by price of temperate product.

cause they are low in absolute or relative terms but because they have been increasing over time slowly and at a diminishing rate. Global scarcity of wood is rising less and less rapidly, which profoundly affects the ability of the tropical timber trade to generate incentives for using tropical timber resources sustainably. This chapter draws heavily on Vincent 1992, but it contains an expanded discussion of the economics of tropical forest management. My perspective is largely Southeast Asian, because Southeast Asia, the dominant exporting region in the tropical timber trade, is where I have the most first-hand experience. Conceptual points, however, are applicable elsewhere.

Economists measure timber scarcity by analyzing the direction and rate of change in stumpage values over time. Expressed volumetrically, stumpage value is the difference between log price and logging cost. It is the maximum net return that a forest owner can collect from harvesting a tree. Like prices of other goods, stumpage values rise if timber becomes more scarce and fall if it becomes less scarce.

In a globally interconnected timber economy, stumpage values reflect global economic timber scarcity. The physical abundance of timber in a given country might have a negligible impact on these values. In particular, if temperate timber is abundant and can easily substitute for tropical timber, then the depletion of old-growth timber in a single tropical country might have little impact on international log prices and therefore on stumpage values in that country.

To see the implications of the divergence between global economic scarcity and physical scarcity within a single country, imagine an underdeveloped tropical country that has not exploited its old-growth forests and cannot import or export wood products. For the moment, assume that forests are valuable only for financial, or market-related, reasons, and that the only significant financial value is provided by timber. In the early stages of economic development, the country would be expected to reduce its area of forest and its stocks of timber rapidly, as it quite rationally (from a financial standpoint) converts forestland to agriculture and uses timber from conversion fellings and additional harvests to build up its infrastructure of railroads, buildings, and so forth. Because timber stocks are large, stumpage values are low.

As timber stocks fall, rising scarcity causes stumpage values to rise. On the supply side, rising stumpage values create an anticipation of higher future returns, which provide an incentive to slow the rate of harvest in old-growth forests and to invest in management of logged forests. On the demand side, rising stumpage values dampen timber consumption by promoting more efficient use of wood during harvesting and processing and increased substitution of nonwood products for wood products. Together, these adjustments promote transition to a sustained-yield state in which harvests equal growth and timber harvests are relatively even over time. In this state, timber is becoming neither more nor less scarce, and so stumpage values are constant.

How does this scenario change if the country is open to trade? Suppose that global timber harvests reduce global timber stocks by a relatively small amount, because harvests are not much greater than growth. Imagine also that the country's forests contain a relatively small proportion of global timber stocks. The first assumption implies that international timber prices, and therefore stumpage values in the country (because it is open to trade), are rising slowly over time. The second assumption implies that depletion of the country's timber stocks will not change this situation.

By definition, net timber growth in old-growth forests is nil, and so growth is not increasing the country's old-growth forests. If stumpage values are not rising significantly, then the forests' value is not increasing because of this factor either. Under these circumstances, the country generates maximum financial value from its old-growth forests by harvesting them as rapidly as possible. It should cash in its old-growth timber at the prevailing international price and invest the realized stumpage value in opportunities that do earn a positive rate of return. In so doing, it converts less productive natural capital to more productive human-made capital. The country's total financial wealth increases in the process.

With the exception of the assumption that timber is the only valuable product of tropical forests, the assumptions of this scenario are not unrealistic. Although world timber demand is rising, there is little evidence that the world is running out of wood. On the supply side, the world has many alternative sources of roundwood for commodity wood products. Roundwood supplies are increasing in many temperate countries both directly, because of increasing areas of plantations and second-growth forests, and indirectly, because of the development of technologies for making reconstituted wood products from mixed species and low-quality timber. On the demand side, increases in global roundwood consumption are diminishing because of three factors: declining population growth rates in nearly all countries; maturing, less resource-intensive economies in developed countries that are growing more slowly; and more intensive use and substitution in response to rising prices. Consumption is rising most rapidly in developing countries, especially for pulp and paper products, which require wood fiber, not high-quality timber.

Binkley and Vincent (1988) compared several leading forest sector models and found that median forecasts were for stumpage values (on a per-m^3 basis) to rise at a real (inflation-adjusted) rate of 2.5%/yr during 1990–2010 but only 1.9%/yr during 2010–2030. This flattening price trajectory is the continuation of a century-long trend. It suggests that, because of market adjustments, global timber scarcity is lessening, not worsening. Sedjo and Lyon (1990) have reached a similar conclusion.

Most if not all tropical countries have a small proportion of global timber stock, because temperate timber dominates the global timber economy. In 1989, tropical countries produced less than a fifth of the world's industrial roundwood (FAO 1991). Timber exports from tropical countries in 1990 accounted for only 11% of the value of global trade in wood products, defined as logs, sawnwood, plywood, and other wood-based panels, pulp, and paper products (Bourke 1992, cited in LEEC 1992).

In sum, global market conditions facing tropical countries correspond closely to theoretical conditions that would induce rapid logging of old-growth tropical forests. This correspondence suggests that, from the standpoint of timber values alone, the rapid logging in most tropical countries since World War II could well be financially optimal (Vincent 1993a). Tropical countries might have suffered long-run financial losses if they had harvested their forests any less rapidly.

Of course, harvest decisions in any forest must take into account not only timber but also other goods and services provided by the forest. Tropical forests provide such nontimber benefits as genetic resources, clean water, recreational opportunities, game, and edible and medicinal plants, to name just a few. These benefits typically lack directly observable market values. Nevertheless, assessing the optimality of rapid logging of old-growth tropical forests from an economic, as opposed to a financial, standpoint requires one to consider the impacts of logging on these benefits.

The economic benefits include future ones about which there is some risk or uncertainty, which economists term option and quasi-option values (Randall 1987). For example, if biological diversity decreases because of logging (see Bennett and Dahaban, this volume), then it is economically rational to reduce the rate of logging to maintain genetic resources until their worth is better known. How much the rate should be decreased is an empirical matter that is difficult to resolve. Panayotou (1992) argues that, given the relative abundance of temperate timber, more or less complete cessation of logging in remaining old-growth tropical forests might be economically justified. In a similar vein, two other economists, Paris and Ruzicka (1991), insist that old-growth forests should be excluded from the portion of a tropical country's forest estate managed for timber production.

SUSTAINABILITY OF TROPICAL TIMBER PRODUCTION

Consideration of the full range of benefits provided by tropical forests leads to the conclusion that rapid old-growth logging is not necessarily economically optimal, even if it is financially optimal. In contrast, determining whether such logging is sustainable would appear to be easy, with the subsequent bust point-ing to an obvious negative answer. The issue of sustainability, however, is per-haps even more complicated than the issue of optimality. Whether one regards rapid old-growth logging as sustainable or unsustainable depends on which of three definitions of *sustainability* one favors.

A traditional forestry definition is "to harvest a country's forests to produce an even flow of timber over time." Under this definition, rapid old-growth logging is obviously not sustainable. Harvests are high for a short pe-riod, but they collapse after all the old-growth forests have been completely logged.

Rapid old-growth logging can be viewed as a sustainable mode of timber production, however, if one relaxes the even-flow constraint. Suppose a tropical country harvests all its old-growth forests in two decades, but it does so in a way that assures regeneration of timber trees. That is, logging is done when there are sufficient seedlings or advanced regeneration (depending on the silvi-cultural system), and these sources of the future timber crop, and the soil on which they depend, are protected from damage during and after logging. If the timber rotation is sixty years, a figure often cited for tropical forests, then the hypothetical country would have no timber to harvest for forty years after the end of the initial timber boom.

Because forest regeneration does occur, however, the country would indeed have a second harvest—in effect, a second timber boom starting sixty years after the first. Harvests would likely be lower during the second boom, given that the forests would have grown for only a few decades, instead of the centuries or millennia involved in the development of old-growth forests. The harvested species almost surely would be different as well (indeed, silvicultural systems typically aim at altering species composition to favor the most valuable species). Nevertheless, this pattern of intermittent or pulse harvesting could continue indefinitely—sustainably—as long as the forests regenerated follow-ing each boom (figs. 14.2 and 14.3).

A second definition of sustainability, simple in concept, though not necessarily in practice, merely requires that forests regenerate following logging. It applies no additional constraints related to the level of harvest over time. The first definition is in fact a special case of the second: it requires that forests not only regenerate but that they be harvested to produce an even flow of timber. The second, regeneration-based definition is thus the purer definition of sus-tainable timber production; the first definition is sustainability with an even-flow constraint. The repeated harvesting of natural (and originally hardwood) forests in the northeastern United States at long intervals, for different types of

FIGURE 14.2   Profile of a selectively logged forest at Jengka Forest Reserve, Pahang, Malaysia. Selective logging removes most overstory trees. Many medium-sized residual trees might die afterward because of logging damage. Photograph by J. Vincent.

FIGURE 14.3   A plantation of kapur (*Dryobalanops aromatica*), a valuable timber tree, established at the Forest Research Institute, Malaysia. Although such systems exist for establishing tropical timber plantations, timber concession policies do not create commercial incentives for investment in such systems.

timber each time, provides a well-documented case of pulse harvesting (Raup 1966). In the tropics, large-scale logging is still a relatively recent phenomenon, but already in Peninsular Malaysia naturally regenerated forests that were originally logged at mid-century have been relogged (S. Appanah, personal communication, 1992; L. C. Cheah, personal communication, 1992). In fact, observations of surprisingly abundant natural regeneration following uncontrolled logging and land degradation during the Japanese occupation of Malaya during World War II provided the basis for the development of the Malayan Uniform System (Appanah and Weinland 1993), which continues to be used successfully to regenerate lowland dipterocarp forests in Peninsular Malaysia.

At a regional level, pulse harvesting might produce a fairly even flow of timber. Since World War II, old-growth logging in Southeast Asia has proceeded, more or less, from the Philippines to Peninsular Malaysia and Sabah to Indonesia and Sarawak and most recently to Papua New Guinea. Although each area logged its old-growth forests rapidly, the entire region's timber production has been maintained at high levels for three decades. Sharp declines in production have already occurred in the Philippines and Sabah. A similar decline will likely have occurred in the remaining areas by the turn of the century. If the forest grows back in each area, however, then the cycle could be repeated starting sometime around the middle of the twenty-first century. In the Philippines and parts of Indonesia, postlogging conversion and degradation of the forest by commercial and subsistence agriculture might severely restrict the forest base for future harvests. But in Malaysia, particularly Peninsular Malaysia, the deforestation rate has fallen sharply since the mid–1970s, as the country's successful industrialization has reduced demand for agricultural land (Vincent and Yusuf 1993). Most of the area covered by forest in Malaysia will still be covered by forest sixty years from now and will be available for a second round of logging.

The International Tropical Timber Organization (ITTO) and environmental groups tend to favor the even-flow definition of sustainability, and they tend to apply it at a national or subnational level (see the report on Sarawak, ITTO 1990c) rather than at a regional level. This definition is explicit in ITTO's "Guidelines for the Sustainable Management of Natural Tropical Forests," which relate sustainability to the forester's notion of an annual allowable cut or coupe (ITTO 1990b). Yet even the most ardent environmentalist might have reason to favor the regeneration-based definition. Testing sustainability by comparing actual harvests or felling areas to allowable ones merely tests whether harvests are uniform during the phase when old-growth timber stocks are being liquidated. It provides no information on the regrowth of forests following logging and thus on the prospects for future harvests in second-growth forests. A country could adhere to perfectly even annual allowable cuts for decades while logging its old-growth forests, but at the same time it might be logging in a way that fails to foster forest regeneration.

Another problem with testing sustainability by the even-flow criterion is that, if it is mechanically applied, it will probably identify virtually all harvesting of old-growth forests as unsustainable, because harvests are inevitably lower in a second-growth forest (Palmer and Synnott 1992: 347). For example, if the timber rotation is sixty years, and, following traditional forestry planning, a country harvests one-sixtieth of its old-growth forest each year, it will nevertheless suffer a decrease in timber production in year sixty-one when it begins harvesting second-growth forests. At the very least, meaningful application of the first definition requires separate calculations of the annual allowable cut during the periods when logging occurs predominantly in old-growth and second-growth forests. Even better, measurements of sustainability should eschew information on the level of harvest as essentially irrelevant and focus instead on the condition of the forest following logging, which is the only source of direct information on the forest's ability to produce a second harvest.

Under both the first and second definitions, *sustainability* does not refer to all aspects of the old-growth forest. By definition, the very act of logging converts the old-growth forest to a second-growth forest, and the ecological changes are perhaps irreversible. Under these two definitions, sustainability refers solely to the forest's ability to produce a commercially valuable timber crop on an ongoing basis. If, even after taking into account nontimber values, rapid old-growth logging is indeed economically as well as financially optimal for a tropical country, then that country would incur economic losses if it chose to harvest its forests in an even-flow manner. If its reason for doing so is international pressure to protect global environmental values, then it has a strong case for requesting compensation from other countries for the values it protects. Otherwise, it is paying the bill for benefits reaped by the rest of the world. Panayotou (1992) discusses issues related to compensation for forest protection.

A third and even less restrictive definition of sustainability looks beyond the forest sector and defines sustainability as "nondeclining, per capita real income (consumption) at the macroeconomic level." According to this definition, even a one-time timber boom associated with the complete elimination of the country's forests' capacity for future timber production can be an integral phase of a sustainable development process. In theory, stumpage value that is invested efficiently can provide fuel for a country's economic takeoff and contribute to diversifying its economy toward industries that rely more on human skills and reproducible capital. By investing stumpage value and other resource rents in human and human-made capital, a country can build an economic base for sustaining its consumption level after the resources run out (Hartwick 1977).

Natural resource accounting (NRA) provides the methodology for determining whether a country's consumption level is sustainable (Dasgupta and Maler 1991; Solow 1992). NRA involves two theoretically equivalent approaches to testing the sustainability of a country's consumption level: (1) calculating a measure of investment that is net of allowances for both depreciation of human-made capital and depletion of natural resources, and checking whether its value is positive; and (2) deducting from gross domestic product (GDP) allowances for both depreciation of human-made capital and depletion of natural resources, and checking whether the resulting net domestic product (NDP) exceeds the country's aggregate consumption level. The first approach is a direct application of concepts in Hartwick 1977, while the second is based on concepts in Weitzman 1976.

The two best-known NRA studies, by the World Resources Institute in Indonesia and Costa Rica, applied only the former approach (Repetto et al. 1989, 1991). They also adjusted GDP, but they did not compare it to consumption. The studies took into account the depletion of not only timber but other resources as well. The study of Indonesia ignored depreciation of human-made capital, apparently owing to lack of data. Both studies found that the adjusted measure of net investment was positive in nearly all years they analyzed (1971–84 in Indonesia and 1970–88 in Costa Rica). Indonesia and Costa Rica apparently accumulated more human-made capital than they depleted natural capital. Their consumption levels therefore appear to be sustainable, in spite of the depletion of their timber and other resources.

Vincent (1993c) applied both approaches to each of the three regions of Malaysia. For Peninsular Malaysia, both approaches indicated that per capita final consumption was sustainable; for Sarawak, only one (the second one); and for Sabah, neither. Hence, findings based on NRA studies indicate that sustainability according to the third definition is being achieved in some, but not all, of the world's principal tropical timber-producing regions.

The third definition of sustainability is surely the one least favored by foresters and environmentalists. But it is not inherently unfriendly to the environment. Under this definition, sustainability requires maintaining permanent forest areas if forests provide environmental services that are essential for the functioning of other economic sectors—for example, watershed services that are essential for agriculture, and genetic resources that are essential for horticultural and pharmaceutical industries. Moreover, when consumption is defined as including not only material goods and services but also nonmarket values—as economists say it should be—then sustainability requires that forests be maintained to provide aesthetic and existence values, recreational opportunities, and other values that contribute to human well-being in a broad sense.

MANAGEMENT OF SECOND-GROWTH TROPICAL FORESTS

The second stage of the boom-and-bust pattern is the lack of management of logged forests (Poore et al. 1989). In practice, management activities can range from actions taken at the time of logging to ensure that regeneration occurs (harvesting when there is an adequate stocking of seedlings, for example, or minimizing damage to the residual stand) to direct investments in postfelling operations (enrichment planting, liberation thinning, and so on). In the absence of management intervention, forests might not grow back at all, thus violating the first and second definitions of sustainability, or they might grow back at suboptimal rates, causing economic losses.

Inefficiencies in tropical stumpage markets bear much of the blame for the lack of management. In fact, most tropical countries have no stumpage markets. Stumpage markets require either private ownership of forests or competitive allocation of logging rights in public forests (through auctioning, for example). Neither arrangement is common in the tropics.

In virtually all tropical countries, forests are government owned, and harvesting is carried out by private parties who receive timber concessions. Concessionaires often have close ties to the government officials who award the concessions. Government forestry departments typically retain responsibility for administering concession agreements, supervising logging, and managing forests before and after logging. Although governments are the owners of the forests, the royalties and other fees they levy on extracted timber typically bear no relation to stumpage values. The fees are set administratively and often somewhat arbitrarily, and they are generally a fraction of stumpage values (Gillis 1980; Page et al. 1976; Repetto and Gillis 1988; Ruzicka 1979; Vincent 1990; Vincent, Awang Noor, and Yusuf 1993). Consequently, most of the stumpage value is captured by concessionaires and their political associates as windfall profits—returns above normal profit margins.

Concessionaires' capture of stumpage value has several implications. First, it suggests that public funds for forest management in many tropical countries are insufficient not so much because of low international timber prices as because of the governments' failure to capture the available stumpage value.

This failure weakens the credibility of tropical countries when they argue—correctly, in theory—that tropical timber boycotts drive down timber prices and reduce the financial viability of forest management (Vincent and Binkley 1992). Actually, tropical countries are not collecting as much revenue as they could.

Second, price-related policies to promote tropical forest management might not achieve their objectives. Some environmental groups, for example, have called for higher tropical timber prices as an incentive for investments in tropical forest management. Even if this were possible—competition with temperate timber would make it difficult—its chief effect would be to raise concessionaires' windfall profits.

Third, most inhabitants of a booming logging region might reap only limited long-run benefits from the timber boom, even if concessionaires diligently invest stumpage value in a way consistent with the third definition of sustainability. Concessionaires are knowledgeable about international investment opportunities and invest their windfall profits in the highest-yielding opportunities. These opportunities are usually in more developed parts of the country, or offshore. This investment pattern means that people without concessions do not derive sustained benefits from either employment, as the timber rents are not invested locally, or investment returns, as they do not share in the windfall timber profits. All they "gain" is a depleted forest. Concessionaires' investments assure their personal financial sustainability, but not the region's economic stability.

The lack of connection between forest ownership and capture of stumpage value hinders both governments and concessionaires from managing second-growth forests. Governments have insufficient financial resources, and concessionaires lack assurance that they will reap the benefits of forestry investments. Concession contracts are typically short, and their renewability is often uncertain. The uncertainty stems from the allocation of concessions as part of a political patronage process in many countries (Hendrix 1990; Kumar 1986; Pura 1990; Repetto and Gillis 1988; Sesser 1991). Concession contracts do not provide the long-term security of private ownership, and the insecurity discourages concessionaires from making long-run forestry investments.

The solution is clear: to combine secure forest tenure with sufficient capture of stumpage value by the party holding the tenure rights (Hyde and Sedjo 1992; Paris and Ruzicka 1991; Vincent 1993b; Vincent and Binkley 1992). Stumpage value from future harvests provides the financial incentive for forest management, and secure tenure provides confidence in this incentive. There are two broad approaches to combining the two. One, the public sector approach, would maintain government ownership and government responsibility for management but would increase the amount of stumpage value captured by the government to a level sufficient for financing viable forestry investments. The other, the private sector approach, would allow concessionaires to continue to capture the lion's share of stumpage value, particularly from future harvests in second-growth forests, but it would restructure concession contracts so that concessionaires have either private ownership rights or rights comparable to those of a private owner. These could be attained by lengthening contracts and making them renewable and transferable, so concessions have asset value.

If tropical countries reform their concession systems to combine forest tenure with capture of stumpage value, then either their forestry departments (under the public sector approach) or concessionaires (under the private sector approach) will need to determine how to manage second-growth forests. This question is largely site specific, but some general remarks can be offered.

First, when stumpage values are rising slowly, the well-known burden of discounting—peoples' preference for economic returns sooner rather than later—remains heavy and forces management interventions to be infrequent and inexpensive (Leslie 1977, 1987). This is especially the case if countries place management responsibility in the hands of concessionaires, whose discount rates are typically higher than the public sector's. Turning the logging operation into the principal management intervention, through careful planning and execution—in effect, scientifically based "log it and leave it"—might offer the best hope for economically viable tropical forest management systems. Elaborate systems of pre- and postlogging treatments may be technically impressive examples of ecosystem manipulation, but they are unlikely to be economically justified.

Second, the global trends toward increasing substitution among timber species and increasing use of reconstituted wood products will force forest managers to rethink the preference for polycyclic selection systems. These trends are making it possible to use an increasing proportion of the species and size classes in tropical forests. Indeed, substitution among tropical timber species has already made the "lesser-known species issue" a nonissue in much of the dipterocarp zone in Southeast Asia. For example, the 1981–82 forest inventory in Peninsular Malaysia found that noncommercial species accounted for only 7.5% of the timber volume in trees with dbh (diameter at breast height, or 1.5 meters above the ground) greater than 30 cm in the region's old-growth forests (Ibu Pejabat Perhutanan, Semenanjung Malaysia, 1987). The minimum-diameter cutting limits under Peninsular Malaysia's Selective Management

System are typically 45–65 cm, but trees as small as 30 cm dbh are indeed harvested and used in areas where minimum-diameter cutting limits are not enforced, such as state-land forests slated for conversion. Violation of cutting limits in forest reserves is emerging as a greater threat to the viability of the Selective Management System than damage to residual trees.

The trend toward fuller use will increase the temptation to violate cutting limits and hamper forestry departments' enforcement efforts. Not only might forestry departments be fighting a futile battle in trying to get concessionaires to abide by cutting limits, but they might also be leaving trees in the forest until they are financially overmature. Cutting limits formulated in the 1970s, when timber markets were more selective in terms of species and log size, might no longer make financial sense, as shown by data from eight concession compartments in Peninsular Malaysia logged in 1988–90 (table 14.4). The estimated stumpage value thirty years later (the length of the cutting cycle under the Selective Management System) assumes, probably optimistically, a gross volume increment of 2.7 m^3/ha/yr and a real rate of stumpage inflation of 2%/yr. Under these assumptions, the rate of return to retaining the residual trees for thirty years ranges between 5.9%/yr and 9.4%/yr. These are respectable rates of return, but they are far below public and private sector discount rates in Malaysia, which are on the order of 13–15% (Veitch 1986).

Harvesting all timber with commercial value at the time of the initial harvest appears to be financially superior to leaving intermediate-sized trees in the forest to grow for thirty years. It would also reduce the temptation for loggers to harvest valuable residual trees illegally by reentering logged compartments before the next cutting cycle. Premature reentry is a major source of damage to regenerating forests in Southeast Asia. Repeated entry into forest cutting blocks damages seedlings, saplings, and small trees of commercial species, reducing future harvests for a small immediate gain.

A principal challenge facing tropical ecologists and silviculturists, therefore, is to develop inexpensive, minimal-intervention management systems that enable forests to regenerate following the removal of virtually all species down to small diameters. Meeting this challenge will probably entail moving away from polycyclic selection systems, which rely on intermediate-sized residuals for the next crop and have short cutting cycles, and toward monocyclic uniform systems, which rely on seedlings, entail longer rotations, and were popular in the tropics until the 1960s and 1970s. Perhaps the best-known tropical silvicultural system, the Malayan Uniform System, was, as its name indicates, a uniform system. Several silviculturists in Malaysia have con-

Table 14.4 Financial analysis of minimum-diameter cutting limits under the selective management system in Peninsular Malaysia

| Compartment | Year 0: Residual timber | | Year 30: Harvested timber | | |
	Timber volume (m³)	Stumpage value (RM/ha)[d]	Timber volume[a] (m³)	Stumpage value[b] (RM/ha)[d]	Return[c] (%/yr)
Pahang 1a	18.5	1,547	64.7	19,571	8.8
Pahang 1b	15.9	1,693	62.1	24,766	9.4
Pahang 2a	43.9	3,156	90.1	23,578	6.9
Pahang 2b	42.9	2,326	89.1	17,667	7.0
Terengganu a	39.3	4,322	85.5	26,826	6.3
Terengganu b	42.3	4,298	88.5	24,273	5.9
Kelantan a	23.6	1,013	69.8	9,877	7.9
Kelantan b	27.8	1,905	74.0	14,559	7.0

Source: Vincent, Awang Noor, and Yusuf 1993.

[a] Assumes gross timber growth rate of 2.7 m³/ha/yr (net rate of 1.54 m³/ha/yr).

[b] Assumes that stumpage value per m³ for timber above the cutting limits rises at 2%/yr.

[c] Rate of increase for values in fourth column compared to those in second column.

[d] RM = Malaysian ringgit (= US $0.37 in 1989).

cluded that that country's experience with both polycyclic and monocyclic systems indicates that monocyclic systems are superior (Appanah and Weinland 1990; Tang 1987).

Intensive logging in tropical forests raises an understandable concern about environmental impacts. Trees as small as 30 cm dbh might be financially mature, but harvesting them could conceivably cause economic losses related to soil erosion, habitat destruction, and other environmental damage. The magnitude of the environmental damage would surely be affected by the size and characteristics of the logged area, its proximity to unlogged areas (directly or via corridors), and the nature of the logging. The relation between logging intensity and environmental damage in tropical forests described by Nussbaum, Anderson, and Spencer (this volume) needs to be incorporated into the economic analysis.

From a theoretical perspective, however, environmental impacts might be a less convincing argument against intensive tropical logging than they first appear to be. The economic theory of comparative advantage states that international trade enables both importers and exporters to make themselves better off through specializing their production. When applied to forest management, this theory predicts that multiple-use management of individual stands is less efficient than dominant-use management: that is, some stands should be managed more intensively for timber production, while others are managed more intensively for production of nontimber values (Vincent and Binkley 1993; see also Swallow and Wear 1993). For any given level of management effort, the theory predicts that the sum value of the output of timber and nontimber values will be greater under dominant-use management than multiple-use management.

This prediction implies that forestry departments should recognize the multiple values of forests under their jurisdiction, but they should manage for production of multiple values at a forest or landscape level, not at a stand level. In stands where the dominant use is timber production, forestry departments should minimize off-site environmental impacts through good logging practices (by prohibiting logging in riparian areas and on too-steep slopes, for example), but managers should not place much emphasis on on-site nontimber values, as these values are being provided more cost effectively by other stands. Dominant use is consistent with the tradition in many tropical countries of dividing a country's permanent forest estate into such management categories as timber production forests, watershed protection forests, and amenity forests. Unfortunately, the economic trade-offs between multiple-use and dominant-use forestry in the tropics have not been carefully analyzed. On the basis of economic and ecological principles and the slim empirical evidence that is available, Panayotou and Ashton (1992) favor dominant use. The most exhaustive empirical examination of the two approaches in temperate forests, by Bowes and Krutilla (1989), found that the dominant-use approach was generally superior.

CONCLUSIONS

Two of the conventional explanations for allegedly unsustainable timber exploitation in the tropics, consumption and import barriers by developed countries, are not well supported by facts. A third explanation, low prices, has more empirical support, but prices are low because of competition with temperate timber, not because of "unfair" market structures. Most important is the slow rate of increase in stumpage values over time, which is an indication that the world is not facing timber shortages. Given this, the rapid logging in old-growth tropical forests and the limited investment in management of second-growth tropical forests could represent financially optimal resource use strategies.

From an economic standpoint, however, the optimality of the boom-and-bust pattern is less clear. Would the sum of timber and nontimber benefits over time be higher or lower if old-growth tropical forests were logged less rapidly? Does environmental damage necessarily rise as tropical logging becomes more intensive (more timber removed per hectare), or do environmental impacts depend more on the size of the logged area, the use of the surrounding land, and the characteristics of the logging operation? Which system provides the greater benefits for a given level of management effort, dominant-use management or multiple-use management?

Of the three definitions of sustainability pertinent to timber production, the most commonly cited one, even-flow timber production, is in fact a special case of a more general definition of sustainable timber production. It amounts to an indirect and potentially misleading criterion for measuring sustainability. Field surveys of regeneration following logging would provide a more direct and reliable means of assessing sustainability.

If dominant-use management is a more cost-effective means of providing both timber and nontimber values, then ecologists' and silviculturists' task in designing sustainable forest management systems becomes simpler. If every stand does not need to provide the full range of goods and services that tropical forests are capable of providing, then management systems can emphasize just one or two products. Testing whether tropical timber is being produced sustainably also becomes simpler. The four pages of "Criteria for Sustainability" issued by ITTA (1990a)—which include such items as "conservation of flora and fauna" and "community consultation"—can be jettisoned in favor of a silvicultural audit to determine whether forests actually regenerate following logging. The audit could be conducted by independent forestry consulting firms, mutually agreed on by exporters and importers and perhaps certified by ITTO or another international body.

Protecting biodiversity and guarding traditional uses of forests are important objectives of forest management, but they are not inextricably linked to the sustainability of timber production. Under the dominant-use approach, these objectives might be better achieved through the management of parks, wildlife reserves, community forests, and other forest areas set aside specifically for the provision of nontimber values. It might be time to shift the focus of the policy debate over the sustainability of tropical timber production from the issue of how to harvest and manage tropical forests at the stand or concession level to the broader issue of how much forest area to allocate to specific use categories.

Appanah, S., and G. Weinland. 1990. Will the management systems for hill dipterocarp forests, stand up? *J. Trop. For. Sci.* 3, 2: 140–158.

———. 1993. Planting quality timber trees in Peninsular Malaysia. Malayan For. Rec. No. 38, Forest Research Institute, Kepong.

Bennett, E. L., and Z. Dahaban. This volume. Wildlife responses to disturbances in Sarawak and their implications for forest management.

Binkley, C. S., and J. R. Vincent. 1988. Timber prices in the U.S. South. *So. J. Appl. For.* 12: 15–18.

Bourke, I. J. 1988. Trade in forest products: A study of the barriers faced by the developing countries. FAO Forestry Paper No. 83, Food and Agriculture Organization, Rome.

———. 1992. Restrictions on trade in tropical timber. Paper prepared for African Forestry and Wildlife Commission, Rwanda.

Bowes, M. D., and J. V. Krutilla. 1989. *Multiple Use Management: The Economics of Public Forestland.* The Johns Hopkins University Press, Baltimore.

Cardellichio, P. A., Y. C. Youn, D. M. Adams, R. W. Joo, and J. T. Chmelik. 1989. A preliminary analysis of timber and timber products production, consumption, trade, and prices in the Pacific Rim until 2000. CINTRAFOR Working Paper No. 22, Center for International Trade in Forest Products, College of Forest Resources, University of Washington, Seattle.

Dasgupta, P., and K. G. Maler. 1991. The environment and emerging development issues. In *Proceedings of the World Bank Annual Conference on Development Economics 1990.* World Bank, Washington, D.C.

Dykstra, D. P., and M. Kallio. 1987a. Base scenario. In *The Global Forest Sector: An Analytical Perspective.* M. Kallio, D. P. Dykstra, and C. S. Binkley, editors. John Wiley & Sons, Chichester, England.

———. 1987b. Scenario variations. In *The Global Forest Sector: An Analytical Perspective.* M. Kallio, D. P. Dykstra, and C. S. Binkley, editors. John Wiley & Sons, Chichester, England.

FAO (Food and Agriculture Organization). 1988. Forest products: World outlook projections. FAO Forestry Paper No. 84. Rome.

——— 1991. *FAO Yearbook of Forest Products: 1978–1989.* Rome.

Gillis, M. 1980. Fiscal and financial issues in tropical hardwood concessions. Development Discussion Paper No. 110, Harvard Institute for International Development, Cambridge.

———. 1988. The logging industry in tropical Asia. In *People of the Tropical Rain Forest.* J. S. Denslow and C. Padoch, editors. University of California Press, Berkeley.

Hartwick, J. M. 1977. Intergenerational equity and the investing of rents from exhaustible resources. *Am. Econ. Rev.* 67: 972–974.

Hendrix, K. 1990. Vanishing forest fells a way of life. *Los Angeles Times,* 18 March.

Hyde, W. F., and R. A. Sedjo. 1992. Managing tropical forests: Reflections on the rent distribution discussion. *Land Econ.* 68, 3: 343–350.

Ibu Pejabat Perhutanan, Semenanjung Malaysia. 1987. *Inventori Hutan Nasional II, Semenanjung Malaysia, 1981–1982.* Kuala Lumpur.

ITTO (International Tropical Timber Organization). 1990a. Criteria for the measurement of sustainable tropical forest management. ITTO Policy Devel. Ser. No. 3, ITTO, Yokohama, Japan.

———. 1990b. ITTO guidelines for the sustainable management of natural tropical forests. ITTO Tech. Ser. No. 5., Yokohama, Japan.

———. 1990c. The promotion of sustainable forest management: A case study in Sarawak, Malaysia. Report submitted to the Eighth Session of the International Tropical Timber Council, May 16–23, Bali.

Kumar, R. 1986. *The Forest Resources of Malaysia: Their Economics and Development.* Oxford University Press, Singapore.

LEEC (London Environmental Economics Centre). 1992. The economic linkages between the international trade in tropical timber and the sustainable management of tropical forests. Draft final report prepared for the International Tropical Timber Organization, under ITTO Activity PCM(IX)/4.

Leslie, A. J. 1977. Where contradictory theory and practice co-exist. *Unasylva* 29, 115: 2–17, 40.

———. 1987. A second look at the economics of natural management systems in tropical mixed forests. *Unasylva* 39, 155: 46–58.

Nussbaum, R., J. Anderson, and T. Spencer. This volume. Effects of selective logging on soil characteristics and growth of planted dipterocarp seedlings in Sabah.

Olechowski, A. 1987. Barriers to trade in wood and wood products. In *The Global Forest Sector: An Analytical Perspective*. M. Kallio, D. P. Dykstra, and C. S. Binkley, editors. John Wiley & Sons, Chichester, England.

Page, J. M., Jr., S. R. Pearson, and H. E. Leland. 1976. Capturing economic rent from Ghanaian timber. *Food Res. Inst. Stud.* 15, 1: 25–51.

Palmer, J., and T. J. Synnott. 1992. The management of natural forests. In *Managing the World's Forests*. N. Sharma, editor. Kendall/Hunt Publishing, Dubuque, Iowa.

Panayotou, T. 1992. Protecting tropical forests. Development Discussion Paper, Harvard Institute for International Development, Cambridge.

Panayotou, T., and P. S. Ashton. 1992. *Not by Timber Alone: Economics and Ecology for Sustaining Tropical Forests*. Island Press, Washington, D.C.

Paris, R., and I. Ruzicka. 1991. Barking up the wrong tree: The role of rent appropriation in sustainable tropical forest management. ADB Environment Office Occasional Paper No. 1., Asian Development Bank, Manila.

Poore, D., P. Burgess, J. Palmer, S. Rietbergen, and T. Synnott. 1989. *No Timber without Trees*. Earthscan, London.

Pura, R. 1990. Battle over forest rights in Sarawak pits ethnic groups against wealthy loggers. *Asian Wall Street Journal*, 26 February.

Randall, A. 1987. *Resource Economics*. Second edition. John Wiley & Sons, New York.

Raup, H. M. 1966. The view from John Sanderson's farm. *For. Hist.* 10, 1: 2–11.

Repetto, R., W. Cruz, R. Solorzano, R. de Camino, R. Woodward, J. Tosi, V. Watson, A. Vasquez, C. Villabos, and J. Jimenez. 1991. *Accounts Overdue: Natural Resource Depreciation in Costa Rica*. World Resources Institute, Washington, D.C.

Repetto, R., and M. Gillis, editors. 1988. *Public Policies and the Misuse of Forest Resources*. Cambridge University Press, Cambridge.

Repetto, R., W. Magrath, M. Wells, C. Beer, and F. Rossini. 1989. *Wasting Assets: Natural Resources in the National Income Accounts*. World Resources Institute, Washington, D.C.

Ruzicka, I. 1979. Rent appropriation in Indonesian logging: East Kalimantan 1972/3–1976/7. *Bull. Indonesian Econ. Stud.* 15, 2: 45–74.

Scott, M. 1989. The disappearing forests. *Far Eastern Econ. Rev.* 143, 12 January: 34–38.

Sedjo, R. A., and K. S. Lyon. 1990. *The Long-Term Adequacy of World Timber Supply*. Resources for the Future, Washington, D.C.

Sesser, S. 1991. Logging the rainforest. *New Yorker*, 27 May: 42–67.

Solow, R. M. 1992. An almost practical step toward sustainability. Invited lecture for the fortieth anniversary of Resources for the Future. Resources for the Future, Washington, D.C.

Sricharatchanya, P. 1989. Too little, too late. *Far Eastern Econ. Rev.* 143, 12 January: 40.

Swallow, S. K., and D. N. Wear. 1993. Spatial interactions in multiple-use forestry and substitution and wealth effects for the single stand. *J. Env. Econ. & Mgmt.* 25, 2: 103–120.

Tang Hon Tat. 1987. Problems and strategies for regenerating dipterocarp forests in Malaysia. In *Natural Management of Tropical Moist Forests*. F. Mergen and J. R. Vincent, editors. Yale University School of Forestry and Environmental Studies, New Haven.

Veitch, M. D. 1986. *National Parameters for Project Appraisal in Malaysia, Volume III: Summary of Estimation Procedures*. Regional Economics Section, Economic Planning Unit, Prime Minister's Department, Kuala Lumpur.

Vincent, J. R. 1990. Rent capture and the feasibility of tropical forest management. *Land Econ.* 66, 2: 212–223.

———. 1992. The tropical timber trade and sustainable development. *Science* 256, 5064: 1651–1655.

———. 1993a. Depletion and degradation are not the same. *J. For.* 91, 4: 24–25.

———. 1993b. Managing tropical forests: Comment. *Land Econ.* 69, 3: 313–318.

———. 1993c. Natural resources and economic growth. Paper presented at the 1993 ISIS Environmental Conference, September 1–2, Kuala Lumpur.

Vincent, J. R., and C. S. Binkley. 1992. Forest-based industrialization: A dynamic perspective. In *Managing the World's Forests.* N. Sharma, editor. Kendall/Hunt Publishing, Dubuque, Iowa.

———. 1993. Efficient multiple-use forestry may require land-use specialization. *Land Econ.* 69, 4: 370–376.

Vincent, J. R., Awang Noor Abd. Ghani, and H. Yusuf. 1993. Economics of timber fees and logging in tropical forest concessions. Harvard Institute for International Development, Cambridge. Manuscript.

Vincent, J. R., and H. Yusuf. 1993. Malaysia. In *Sustainable Agriculture and the Environment in the Humid Tropics.* National Research Council, editor. National Academy Press, Washington, D.C.

Weitzman, M. L. 1976. On the welfare significance of national product in a dynamic economy. *Q. J. Econ.* 90, 1: 156–162.

          R. A. Houghton

What is the value of a tropical forest? This is not one question but two. The first is a scientific question: what are the goods and services provided by forests? Goods and services include not only timber and nontimber products but also the services forests provide in stabilizing flows of water, improving water quality, modifying the chemistry of the atmosphere, and affecting climate. The second question is economic: what is the monetary value of a tropical forest? The scientific answer must precede the economic one if the true value of forests is to be determined. Because forests are being used as though their goods and services were known and worth little, a scientific assessment of the goods and services provided by forests is urgently needed.

Managed sustainably, forests can indefinitely supply timber, fuel, and nontimber products, as well as genetic resources and ecosystem services. Trends in much of the tropics assume that sustainably managed forests are worth less than the short-term value of the wood and land they hold; forests are being mined at a rate that will exhaust them in less than a century. During the twentieth century about 25% of the area of tropical forests was converted to agriculture, and the rate of conversion increases yearly (Houghton 1994). If commodities extracted from forests are worth more than the forests themselves, there is no incentive to use forests renewably.

At the United Nations Conference on Environment and Development (UNCED) in Rio de Janeiro in 1992, international conventions on climate and biodiversity were accepted by most countries. A convention on forests was not accepted, although a nonbinding statement of principles was created. A convention on forests would have been premature because an evaluation of current knowledge about the world's forests, including an assessment of the goods and services they provide (Ramakrishna and Woodwell 1991, 1993), is a prerequisite for such a convention. The assessment of what scientists know about climatic change, for example, was comprehensively documented by the scientific community under the auspices of the Intergovernmental Panel on Climate Change (IPCC) (Houghton, Callander, and Varney 1992; Houghton, Jenkins, and Ephraums 1990). The IPCC was established in 1988 by two United Nations organizations, the World Meteorological Organization (WMO) and the United Nations Environment Programme (UNEP). The scientific assessment took two years and involved almost four hundred scientists from twenty-five countries.

Other chapters in this volume document the importance of forests to national economies, local cultures, and species diversity. Forests also provide goods and services that might, if identified, lead to a greater valuation of forests and, perhaps, to their preservation. This chapter documents the importance of forests to climate and climatic change and summarizes what is known about global climatic change. Although the major causes of climatic change have been and continue to be industrial activities, this chapter will emphasize the contributions of forests. The bias is appropriate for a book on tropical forests but is not meant to imply that management of tropical forests should be used to offset the bad habits of the industrial world. Stabilizing global climate is not the only reason for preserving and managing forests. Local and regional effects of deforestation include erosion of soil, increased frequency of both floods and droughts, and other changes in climate that affect both people and forests adversely. These deleterious effects of deforestation help define the goods and services provided by forests.

GLOBAL CLIMATIC CHANGE

Scientists understand some elements of global climatic change, but other elements are controversial because not enough data are available to decide between alternative interpretations or to test predictions. Established facts are (1) that the "greenhouse effect" is real and (2) that the gases responsible for this effect are increasing in the atmosphere as a result of human activities.

Measurements made by satellites confirm that the atmosphere is more transparent to short-wave (visible) light received from the sun than to longer-wave (infrared or heat) radiation from the earth. As a result, light can enter the atmosphere more easily than heat can escape; under normal conditions, the earth's atmosphere heats until outgoing long-wave radiation equals incoming short-wave radiation. Calculations show that, as a result of this natural greenhouse effect, the earth is about 33°C warmer than it would be otherwise. Thus, the greenhouse effect is a good thing. Without it, the average temperature of the earth's surface would be below 0°C, and life would not exist.

The greenhouse effect is a fact. It is no more a hypothesis than is gravity. Concern about the greenhouse effect is, strictly speaking, a concern about the *enhanced* greenhouse effect expected as a result of emissions of greenhouse gases to the atmosphere.

The second undisputed fact is that concentrations of various greenhouse gases (carbon dioxide, methane, nitrous oxide, and chlorofluorocarbons) in the atmosphere are increasing as a result of increased emissions of these gases from human activities, primarily combustion of fossil fuels. The increased concentrations will enhance the greenhouse effect and hence cause the surface of the earth to warm. Concentrations of the principal greenhouse gas (water vapor), which are not under human control, will increase as a result of the warming. How much the earth will warm, and how rapidly, depends on the rates at which humans continue to emit greenhouse gases to the atmosphere.

Not all scientists agree about the warming already observed. The mean global surface air temperature is generally thought to have risen about 0.3°C–0.6°C since 1880 (Houghton, Jenkins, and Ephraums 1990). The five warmest years before 1990 occurred in the 1980s; 1990 and 1991 were the warmest years on record. Some scientists, however, argue that the "observed" warming may be an artifact resulting from the small number of sampling stations in the early part of the record, a biased distribution of stations around the world, local heating effects on some stations, and systematic changes in the way temperature is measured. For example, canvas buckets used to measure temperature at sea were gradually replaced by metal buckets in the first half of the twentieth century. Each of these uncertainties has been addressed by the scientists who analyzed and published the temperature records, but the apparent warming is not universally accepted in the scientific community.

The largest uncertainty in the study of climatic change is whether the increased concentrations of greenhouse gases caused the observed warming. There is no scientific proof, and, indeed, there may never be proof, that the two phenomena are causally linked. The warming may be the result of natural phenomena, such as variability in the sun's output. Indeed, the warming has not been consistent. Between 1940 and 1970, for example, the earth cooled, though atmospheric concentrations of greenhouse gases increased.

Scientists are generally not interested in what is known; the important business of science is investigation of the unknown or uncertain. This emphasis on uncertainties tends to obscure the elements of the climate system about which there is general agreement and promotes resistance to action addressing climate change. Short-term economic and political interests also tend to resist change. One form of resistance is to deny the likelihood of climatic change and to focus on the uncertainties; however, uncertainty works in both directions. True, global warming may not be as rapid or as large as the models suggest, but it is equally likely to be larger and more rapid than predicted.

The Relative Effectiveness of Greenhouse Gases

The properties of the major greenhouse gases are known well enough for scientists to calculate which ones are likely to be most important in the warming of the earth. The calculation depends on three attributes of each gas: its radiative properties, its atmospheric lifetime, and its emission rate. The radiative properties of greenhouse gases depend on their molecular structures. The radiative strength, or forcing, of each gas varies. One kg of methane is about 35 times more effective in heating the earth than 1 kg of CO_2, 1 kg of N_2O is about 260 times more effective than 1 kg of CO_2, and 1 kg of chlorofluorocarbons (CFCs) is thousands of times more effective than 1 kg of CO_2 (table 15.1). Each of the many CFCs has a different radiative forcing; 1 kg of the major CFCs is 4,000 to 7,000 times more effective than 1 kg of CO_2.

The second relevant property of greenhouse gases is their atmospheric lifetime, the average time a molecule remains in the atmosphere before breaking down or leaving the atmosphere. Atmospheric lifetimes are determined from an understanding of the quantities of the gas in the atmosphere and rates of emission or breakdown. The lifetimes of the major greenhouse gases, except for methane, are long. A molecule of nitrous oxide emitted this year will remain in the atmosphere for about 130 years. Again, the lifetimes of CFCs vary widely. The atmospheric lifetime of CO_2 is uncertain, largely because estimates of carbon emissions do not currently balance estimates of carbon accumulations (see table 15.1). Because lifetimes of gases differ, the relative effectiveness of a gas in trapping heat depends, in part, on the time frame of interest. If the interval in question is short, the relative effect of a short-lived gas (such as methane) is greater.

The third aspect determining the effectiveness of a greenhouse gas is the rate at which it is emitted to the atmosphere. Emissions are clearly dominated by CO_2 and, indeed, over the next hundred years CO_2 is calculated to account for about 60% of the warming expected in the IPCC business-as-usual scenario (Houghton, Jenkins, and Ephraums 1990). This scenario assumes that current growth rates in global energy use will continue, although in practice the production of most CFCs will be phased out.

Knowing the radiative properties, the atmospheric lifetimes, and the emission rates of various gases allows scientists to calculate not only the relative contributions of different gases to future warming but also the reductions in emissions required to stabilize concentrations of the gases in the atmosphere.

Table 15.1 The characteristics of greenhouse gases and their relative contributions to a predicted global warming over the next 100 years

Gas	Warming effect of 1 kg relative to that of CO_2	Atmospheric lifetime (years)	Atmospheric emissions in 1990 (MMT)	Relative contribution over 100 years (%)
Carbon dioxide	1	50–200[a]	26,000[b]	61
Methane	35[c]	10	300	15
Nitrous oxide	260	130	6	4
CFCs and HCFCs[d]	1,000s	10s to 100s	1	11
Others[e]				9

Source: Houghton, Jenkins, and Ephraums 1990.

[a] The broad range in lifetime for CO_2 results from uncertainties in the global carbon cycle and from the fact that the removal of CO_2 from the atmosphere depends not only on the amount but on the rate at which CO_2 is emitted to the atmosphere.

[b] 26,000 MMT CO_2 = 7 BMT carbon (MMT = million metric tons; BMT = billion metric tons). 7 BMT carbon includes 6 BMT from combustion of fossil fuels and 1 BMT from deforestation. Emissions of carbon from changes in land use are estimated to have been about 2 BMT in 1990.

[c] Includes the indirect effects to concentrations of other greenhouse gases through chemical interactions in the atmosphere.

[d] The radiative properties and atmospheric lifetimes of specific CFCs and HCFCs are known precisely. Only order-of-magnitude averages for the more abundant gases are shown here.

[e] Principally tropospheric ozone, indirectly generated in the atmosphere as a result of other emissions.

Warming will continue as long as the concentrations of these gases in the atmosphere increase. The concentrations, in turn, will continue to increase as long as the gases are emitted to the atmosphere at current rates, or even at very reduced rates. According to Houghton, Jenkins, and Ephraums 1990, stabilization of concentrations at present-day levels would require reductions in emission rates of 60% for the long-lived gases (all but methane).

This point is important. The most progressive industrial nations are committing themselves to 25% reductions in emissions of carbon dioxide by early in the twenty-first century. Some countries, such as the United States, have resisted any reductions in carbon dioxide; they would prefer to continue business-as-usual and allow emissions to increase. Even the most progressive reductions, however, are considerably less than the 60% required for stabilization. This difference between political and scientific views is remarkable. As long as the political view prevails, concentrations will continue to increase for years. As Houghton, Jenkins, and Ephraums (1990) point out, stabilization of future concentrations, after emission rates have been allowed to increase, will require even larger reductions. Without strong, deliberate, immediate action, the world seems destined for warming and for other changes in climate more difficult to predict. Even with immediate emissions reductions on the order of 60%, the earth is committed to a warming of another 0.5°C beyond what has already been observed.

Rates of Emission of Carbon Dioxide

Of the greenhouse gasses emitted because of human activities, CO_2 is expected to contribute most to global warming. Carbon dioxide is emitted to the atmosphere as a result of industrial activity (combustion of fossil fuels) and large-scale land-use change (deforestation). Combustion of fossil fuels started during the industrial revolution in the mid-nineteenth century and has increased almost continuously since that time (fig. 15.1). Before 1850, global emissions were essentially zero; in 1990 emissions of carbon to the atmosphere were 6 billion metric tons (BMT) (or 22 BMT CO_2). The only interruptions to the exponential increase in emissions occurred during World Wars I and II and the Great Depression, and after the abrupt increases in the price of oil during the 1970s and 1980s. Use of fossil fuels, and hence the annual emission of CO_2 to the atmosphere, increased each year until 1979, except during the World Wars. As a result of price changes and a weakening of the global economy, annual emissions between 1979 and 1983 decreased each year, but the downward trend was temporary. After 1983, annual emissions increased again, so that by 1985 they

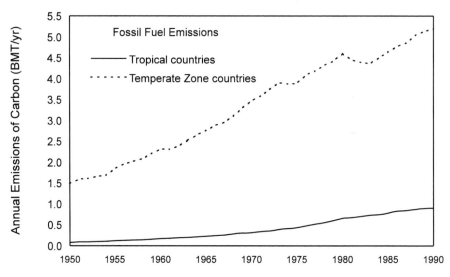

FIGURE 15.1   Annual emissions of carbon from combustion of fossil fuels. *BMT* = billion metric tons.

were similar to what they had been in 1979. Since 1985, global emissions have increased each year. The total emissions of carbon to the atmosphere from fossil fuels have been about 200 BMT since the start of the industrial revolution.

The relative contributions of regions to the annual emissions of fossil fuel carbon are changing. In 1925, the United States, Western Europe, Japan, and Australia were responsible for about 88% of the world's fossil fuel CO_2 emissions (Darmstadter 1971). By 1950 the fraction contributed by these countries had decreased to 71%, and by 1980, to 48% (Rotty and Marland 1986). The annual rate of growth in the use of fossil fuels in these countries varied between 0.5% and 1.4% in the 1970s. In contrast, the annual rate of growth in fossil fuel use in the developing nations remained at 6.3% per year during the 1970s. The developing countries' share of the world's fossil fuel consumption grew from 6% in 1925 to 10% in 1950 to about 20% in 1980. By 2020, the developing world is projected to consume annually about 60% of the world's fossil fuels (Goldemberg et al. 1985).

BIOTIC CARBON FROM CHANGES IN LAND USE

Changes in land use transfer carbon between land and atmosphere in either direction. Deforestation—the conversion of forests to cleared lands—releases carbon to the atmosphere through burning and decay of trees and soil (fig. 15.2). In contrast, regrowth of forests withdraws carbon from the atmosphere and stores it in trees and soil organic matter. The long-term trend in the net flux of carbon from both deforestation and reforestation has been an increasing release of carbon as the area of the world's forests has been reduced (fig. 15.3). The trend in the net release is similar to that of fossil fuels, with two important differences. First, current emissions from changes in land use are considerably less (about 2 BMT) than current emissions from combustion of fossil fuels (6 BMT), and second, the major contributors to deforestation are the less developed countries of the tropics. In 1985 the net release of carbon from deforestation came almost entirely from the tropics, whereas the emissions of CO_2 from fossil fuels came largely from outside the tropics (Houghton et al. 1987).

Tropical countries did not always emit more carbon from changes in land use than temperate-zone countries. The net release of biotic carbon from the tropics was relatively unimportant until about 1950 (see fig. 15.3). In the nineteenth century the major regions of biotic flux were the industrialized regions of North America, Europe, and Russia, with some contribution from densely populated regions of South Asia and China.

The net annual biotic flux was probably less than 0.5 BMT before 1850 and probably less than 1 BMT carbon until 1950. The combined biotic and

FIGURE 15.2   A tropical rainforest was removed from this land near Bintulu, Sarawak, to establish an oil palm plantation. Oil palm plants, approximately 1 m in height, have been planted. Photograph by R. Primack.

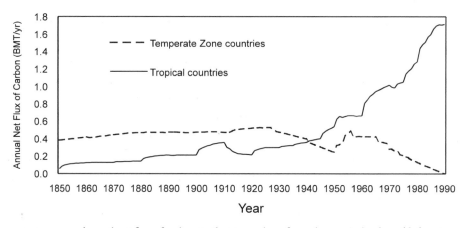

FIGURE 15.3   Annual net flux of carbon to the atmosphere from changes in land use (deforestation and reforestation).

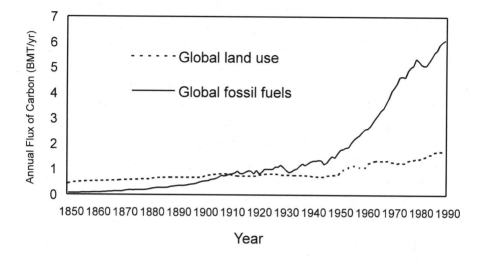

FIGURE 15.4   Annual global emissions of carbon from combustion of fossil fuels and changes in land use.

fossil fuel sources first exceeded 1 BMT/yr near the start of the twentieth century (fig. 15.4). It was not until about 1910 that the annual emissions of carbon from combustion of fossil fuels exceeded the net biotic flux. Since 1950, the fossil fuel contribution has predominated, although emissions from both sources have increased with the intensification of industrial activity and the expansion of agricultural area. The net flux in 1990 was 7 to 8 BMT carbon per year: 6.0 BMT from fossil fuels and 1.5 to 2.0 BMT from biotic sources.

The switch to fossil fuels did not noticeably reduce emissions from wood fuels (see fig. 15.4). One reason the switch is not evident is that it occurred in different regions of the world at different times. It has not yet occurred in many of the developing nations. In Africa and South Asia, for example, fuelwood still accounts for more than 80% of total fuel use. The major reason why increases in fossil fuel emissions were not accompanied by decreasing biotic emissions is that most of the biotic carbon was released not from wood fuels but from the oxidation of vegetation and soils associated with the expansion of cultivated land. The harvest of forests for fuelwood is not important in adding carbon to the atmosphere because the emissions from combustion are usually balanced by the accumulation of carbon in regrowing forests. The balance will occur only as long as the forests harvested for fuelwood are allowed to regrow, however. If the gathering of fuelwood leads to permanent deforestation, the process will result in a net release of carbon to the atmosphere.

Calculation of the net flux of carbon from changes in land use is based on two types of information, rates of land-use change and changes in the vegetation and soils of the affected ecosystems. Forests contain twenty to fifty times more carbon in their vegetation than the agricultural lands that replace them. Rates of land-use change can therefore be determined from information about the area of forests—or, lacking that, the area of croplands, pastures, and other managed lands—through time. In tropical Asia the expansion of croplands and shifting cultivation seems to have been responsible for almost all the deforestation in the region. The rate of deforestation has increased throughout the twentieth century (Houghton and Hackler 1994; Richards and Flint 1994). In the 1980s the rate increased by almost 100%, according to the Food and Agriculture Organization (FAO 1993). Myers (1991) independently estimated a 78% increase over the same period, and he believes that this figure represents a real increase in rates between 1979 and 1989. In contrast, the FAO (1993) acknowledges that some of the difference in its estimates resulted from underestimating the rate of deforestation in earlier years (fig. 15.5). The FAO recognizes that deforestation has accelerated in tropical moist regions as a whole, although in some Asian countries it has declined. The revision in the rates for 1980 brings the FAO's estimate more in line with Myers's earlier estimate.

Different kinds of forests hold different amounts of carbon in their

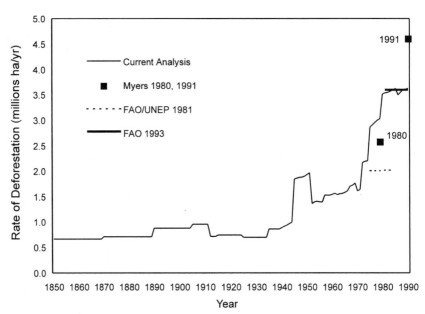

FIGURE 15.5 Annual rates of deforestation in South and Southeast Asia (from Houghton and Hackler 1994).

biomass and soils. For the forests of South and Southeast Asia, two estimates of carbon in biomass are available for each type of forest (table 15.2). The higher estimates are based on direct measurement of biomass (Bruenig 1974; Edwards and Grubb 1977; Hozumi, Yoda, and Kira 1969; Kira 1978; Ogawa et al. 1965; Proctor et al. 1983; Sabhasri 1978). The lower estimates are determined from volumes of wood (FAO/UNEP 1981). Based on the low estimates, the weighted average for closed forests (moist and seasonal forests) is 116 tC/ha (metric tons carbon per hectare), intermediate between the weighted averages (92 and 130 tC/ha) calculated by Brown, Gillespie, and Lugo (1989 and 1991, respectively) for South and Southeast Asia. Brown, Gillespie, and Lugo's estimates of aboveground biomass were converted here to total carbon stocks, assuming belowground biomass to be 16% of aboveground biomass (Brown, Gillespie, and Lugo 1989) and assuming biomass to be 50% carbon.

The carbon content of soils for these ecosystems was derived from data published by Yoda and Kira (1969); Schlesinger (1977); Zinke, Sabhasri, and Kunstadter (1978); Brown and Lugo (1982); and Post et al. (1982) (see table 15.2). Reviews by Detwiler (1986) and Schlesinger (1986) indicate that clearing of a forest and subsequent cultivation of the soils reduces the carbon contained in the top meter of soil by about 25%, although this figure varies. Most of this loss occurs within the first five years after clearing.

When forests are cleared for croplands, some of the original biomass is burned, some is left as "slash" to decay on the surface or belowground, and some may be removed for wood products, which are slow to oxidize. (Slash refers to the stumps, roots, and unburned woody debris left on site after clearing and burning.) The net loss of carbon from deforestation depends on the management of the deforested land. If rubber or palm oil plantations are established, for example, the net loss of carbon will be less than if the land were cleared for annual crops (table 15.3).

Table 15.2 Carbon in the vegetation and soils of forest ecosystems in tropical Asia

	Moist forests (metric tons/ ha of carbon)	Seasonal forests (metric tons/ ha of carbon)	Dry forests (metric tons/ ha of carbon)
Carbon in vegetation			
High estimates of biomass	250	150	60
Low estimates of biomass	135	90	40
Carbon in soils	120	80	50

Source: Palm et al. 1986.

Table 15.3 Percentage of initial carbon stocks lost to the atmosphere when tropical forests are converted to different land uses.

Land Use	Vegetation	Soil
Cultivated land	90–100	25
Pasture	90–100	12
Degraded croplands and pastures[a]	60–90	12–25
Shifting cultivation	60	10
Degraded forests	25–50	< 10
Logging[b]	10–50	< 10
Plantations[c]	30–50	< 10
Extractive reserves	0	0

Source: Data from Houghton et al. 1987 unless otherwise indicated.

Note: For soils, the stocks are to a depth of 1 m. The loss of carbon may occur within 1 year, with burning, or over 100 years or more, with some wood products.

[a] Croplands and pastures, abandoned because of reduced fertility, may accumulate carbon, but their stocks remain lower than the initial forests.

[b] Based on estimates of aboveground biomass in undisturbed and logged tropical forests (Brown, Gillespie, and Lugo 1989). When logged forests are colonized by settlers, the losses are equivalent to those associated with one of the agricultural uses of land.

[c] Plantations may hold as much carbon as natural forests or more, but a managed plantation will hold, on average, 1/3 to 1/2 as much carbon as an undisturbed forest because it is generally regrowing from harvest.

Logging is not considered deforestation because logging does not reduce the area of forests unless logged areas are subsequently colonized and farmed. Logging does reduce the amount of carbon held in forests, however, unless logged forests are given time to recover (see table 15.3). These reductions in biomass are much harder to quantify than reductions in area, but they seem to be widespread throughout the tropics. Indeed, an FAO survey (1993) reported that the loss of biomass in tropical forests was occurring at a significantly higher rate than the loss of forest area.

Two independent studies produced some of the first quantitative estimates of this loss of biomass from within forests. Brown, Gillespie, and Lugo (1991), in a study based on surveys of Peninsular Malaysia by the FAO (1973) and the Government of Malaysia (1987), found that an 18% reduction in forest area coincided with a 28% reduction in total biomass. The total loss of biomass was 267.6 MMT (million metric tons) carbon, and the total loss of forest area was 1.45 million ha. Thus, the ratio of biomass lost to area lost was 184.5 tons carbon per hectare (t C/ha). In contrast, the average biomass of these forests at the start of the interval was 115 t C/ha. Thus, the average carbon lost was 1.6 times larger per unit area than the average biomass of the initial forests. A possible interpretation of this ratio is that for every ton of carbon released to the atmosphere through deforestation, an additional 0.6 ton of carbon was released from the reduction of biomass in the remaining forests. Such degradation appeared to occur in all types of forests.

In a similar study, Flint and Richards (1994) estimated that for almost all of South and Southeast Asia a 34% loss of forest area between 1880 and 1980 coincided with a 55% loss of forest biomass in the same period. The loss of biomass was 0.62 times greater than the loss that could be explained by deforestation. The study also found that the relative importance of biomass reduction had not changed in a hundred years. This surprising fact indicates that loss of biomass did not become less important as such practices as shifting cultivation were displaced by permanent cultivation. As more forests were cleared for agricultural purposes, logging and fuelwood harvest increased proportionally.

The reduction in biomass as a result of selective harvest began well before the twentieth century. According to Flint and Richards (1994), the average biomass of South and Southeast Asian forests in 1850 was already about 25% less than the biomass of undisturbed forests. This amount of degradation is not surprising in a densely populated area. After 1850 degradation accelerated, and in 1990 biomass averaged about 60% of its assumed undisturbed value. Further, the area of forest in South and Southeast Asia decreased by 38% (176 million ha) over this 140-year period. Together, changes in area and changes in biomass within forests appear to have reduced the storage of carbon in the vegetation of tropical Asia by about 50%.

The net flux of carbon calculated to have been released to the atmosphere from South and Southeast Asia because of changes in the area of forests and changes in biomass within them increased from about 0.1 BMT carbon/yr^{-1} in the mid-nineteenth century to about 0.7 BMT carbon/yr^{-1} in 1990 (fig. 15.6). The total net release between 1850 and 1990 was 29.5 BMT (Houghton and Hackler 1994). Deforestation accounted for 67% (20 BMT carbon) of the flux; 33% (10 BMT) was from reductions in biomass as a result of selective logging. Soils accounted for about 15% of the flux (4.5 BMT). Another 6 BMT carbon accumulated in long-term storage products (such as wood houses and furniture) removed from forests. If the carbon in those products had been released to the atmosphere, the total net flux would have been about 35 BMT carbon.

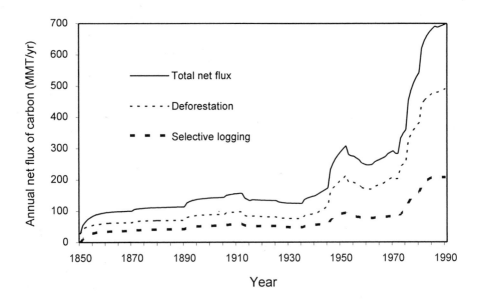

FIGURE 15.6 Annual net flux of carbon to the atmosphere from deforestation and selective logging in South and Southeast Asia (from Houghton and Hackler 1994). *MMT* = million metric tons.

EMISSIONS OF CARBON FROM OTHER TROPICAL LEGIONS

Estimates of the flux of carbon from changes in land use worldwide vary between 0.6 and 2.5 BMT carbon for 1980 (Hall and Uhlig 1991; Houghton 1991; Houghton et al. 1987) and between 1.1 and 3.6 BMT carbon for 1990 (Houghton 1991). Almost entirely from the tropics, this net release of carbon includes the accumulation of carbon in forests regrowing after logging and after abandonment of agriculture.

These published estimates need revision. The high estimate for 1990 was based on an apparent increase in the rate of deforestation in the Brazilian Amazon (Myers 1991). Myers based his estimate of Brazilian deforestation on a study by Setzer and Pereira (1991), who used Advanced Very High Resolution Radiometer (AVHRR) data from the NOAA–7 satellite to determine the number of fires burning in Legal Amazonia (an area of 500 million ha that includes most of the Brazilian Amazon) during the dry season (mid-July through September). Setzer and Pereira recognized that some fires burned for more than one day (and should not be counted twice) and that a small, hot fire would saturate the entire pixel (1 km²) and overestimate the area actually burned. With these adjustments, they estimated that about 20 million ha of fires burned in the Brazilian Amazon in 1987, about 60% (12 million ha) of which were on lands that had already been deforested. Their estimate of deforestation was 8 million ha.

Myers (1991) reduced the estimate by Setzer and Pereira to 5 million ha, to account for other factors. Nevertheless, even this reduced rate seems high according to more recent studies. Using data from Landsat (80–m resolution rather than 1–km resolution), the Brazilian National Institute for Space Research (INPE) found the rate of deforestation of closed forests in Brazil's Legal Amazonia to have averaged about 2.1 million ha/yr between 1978 and 1989 (Fearnside, Tardin, and Filho 1990), about one-fourth the rate initially determined by Setzer and Pereira (1991). Skole and Tucker (1993) reported an average rate of 1.5 million ha/yr for almost the same interval. The actual rate probably increased between 1978 and 1987 but fell substantially after 1987 to 1.8 million ha in 1988–89, 1.38 million ha in 1989–90, and 1.11 million ha in 1991.

When the more recent data from INPE are substituted for those reported by Myers (1991), the estimated rate of deforestation for Latin America is revised from 7.7 to 4.5 million ha/yr. The rate of deforestation throughout the humid tropics appears to have increased by about 40% rather than 90% between 1980 and 1990. The revised estimated rate for the late 1980s is 10.66 million ha/yr for closed forests alone. The FAO (1993) estimates the total rate for closed and open tropical forests at 17.1 million ha/yr.

These revised rates of deforestation along with data from Houghton, Skole, and Lefkowitz 1991 were used to recalculate emissions of carbon from Latin America. The effect of the revisions was to reduce the calculated emissions from about 0.7 BMT C/yr (Houghton, Skole, and Lefkowitz 1991) to about 0.5 BMT C/yr in 1980, and from about 0.9 to 0.7 BMT C/yr in 1990 (fig. 15.7). Error is thought to be within ± 50%. The data used to calculate the flux of carbon from tropical Asia and Africa have also been revised. Rates of deforestation in Southeast Asia were revised upward by the FAO (1993; see fig. 15.5), and Flint and Richards (1994) reassessed historical rates of degradation in the region. The revised estimate of flux for South and Southeast Asia was about 0.7 (± 0.3) BMT carbon in 1990 (Houghton and Hackler 1994). A reanalysis of land-use change in Africa, using revised estimates of biomass (Brown, Gillespie, and Lugo 1989), produced an estimate of flux there of 0.35 (± 0.2) BMT carbon in 1990 (see fig. 15.7).

As a result of these revisions, the global net flux is estimated to have been about 1.4 BMT carbon in 1980 (1.3 BMT from the tropics and 0.1 BMT from outside the tropics) and 1.7 (± 0.5) BMT carbon in 1990 (essentially all from the tropics). The cumulative flux for 1850–1990 was 120 BMT carbon (see fig. 15.4).

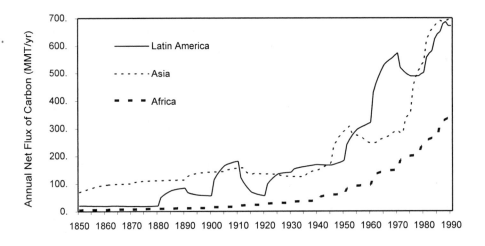

FIGURE 15.7   Annual net flux of carbon to the atmosphere from changes in land use in tropical Asia, tropical America, and tropical Africa. *MMT* = million metric tons.

DEFORESTATION'S CONTRIBUTION TO GLOBAL WARMING

Carbon dioxide is not the only form of carbon emitted as a result of changes in land use. Other forms of carbon, such as carbon monoxide and methane, are released to the atmosphere when forests or fields are burned. About 10% of the carbon may be released in the form of carbon monoxide (more if the fire is a slow, smoldering one, less if the fire is burning vigorously with open flames). Although carbon monoxide is not a greenhouse gas, it indirectly affects the concentrations of methane by its chemical reactions with the hydroxyl radical in the atmosphere. The hydroxyl radical is destroyed by carbon monoxide, methane, and other gases, so increased emissions of any of these gases can reduce atmospheric concentrations of the hydroxyl radical and thereby prolong the lifetime of other gases.

About 1% of the carbon produced by burning of vegetation may be emitted as methane. In addition, methane is emitted from anaerobic environments, such as rice paddies, and the enteric fermentation of animals (especially ruminants). Another greenhouse gas, nitrous oxide (N_2O), is also emitted through burning, but the major source of atmospheric N_2O is the soils of recently cleared pastures and fertilized croplands. Each of these gases has industrial sources as well as biotic sources. The chlorofluorocarbons, however, are entirely of industrial origin. As already discussed, the effectiveness of these greenhouse gases can be calculated from their radiative properties, atmospheric lifetimes, and emission rates. Calculations for the 1980s showed CO_2 to be most important, followed by CFCs, methane, and finally N_2O (Houghton, Jenkins, and Ephraums 1990). Houghton, Jenkins, and Ephraums (1990) have determined the proportion of each gas that is emitted from industrial and biotic sources, as described here for CO_2. The information can be summarized to indicate the relative importance of industrial activities and land-use change for global warming (table 15.4). The contribution to global warming from changes in land use, largely tropical deforestation, is 20–25%.

Table 15.4 Relative contribution of tropical deforestation and subsequent use of land to the greenhouse effect calculated for the 1980s

	Annual emissions	Percentage of total emissions	Calculated contribution to radiative forcing	
			Total[a]	Tropical deforestation
Carbon dioxide	10^9 tons carbon		55%	
Industrial	5.6			
Natural	0.0			
Deforestation	2.0	25%		14%
Total	7.6			
Methane	10^6 tons CH_4		15%	
Industrial	100			
Natural	150			
Deforestation	250	50%		7.5%
Total	500			
Nitrous oxide	10^6 tons N_2O		6%	
Industrial	1.3			
Natural	7.2			
Deforestation	2.0	20%		1.0%
Total	10.5			
CFCs and HCFCs	10^3 tons CFC		24%	
Industrial	1.0			
Natural	0.0			
Deforestation	0.0	0%		0.0%
Total	1.0			
Total			100%	20–25%

[a] From Houghton, Jenkins, and Ephraums 1990

If the current rates of deforestation in the tropics continue for another hundred years, most tropical forests will disappear and approximately 100 to 300 BMT of carbon will be released to the atmosphere (Houghton 1990). In comparison, about 100 BMT of carbon have been released from changes in land use since the mid-nineteenth century, and about 300 BMT of carbon have been released to date from worldwide changes in land use and from combustion of fossil fuels (see fig. 15.4). Few people believe that tropical forests will be eliminated, but the current rate of deforestation reduces the area of remaining forests by about 1% each year, and the rate seems to have increased by about 40% in the 1980s. The FAO interim report (1993) suggested that where rates of deforestation had been lowered during this period, the reduction was caused by the loss of forests rather than by deliberate decisions to preserve the remaining stands.

From the perspective of climate, forests are important. They may be managed to increase or reduce the rate of global warming. The major contributor to warming, however, is industrial emissions of greenhouse gases, largely from countries outside the tropics. Those countries must act to reduce global warming. Tropical countries are responsible for comparatively little of the accumulation of greenhouse gases in the atmosphere, but their relative contribution is increasing, owing to their growing use of fossil fuels (see fig. 15.1) and their increasing rates of deforestation (see fig. 15.3).

Several possible management strategies illustrate the potential role of forests in stabilizing concentrations of atmospheric CO_2 (table 15.5). The first strategy is what is currently practiced today. About 8 BMT carbon are released to the atmosphere annually, 6 BMT from fossil fuels and 2 BMT from deforestation. Reforestation of cleared land is not globally significant.

The second strategy, massive reforestation through either natural regeneration or planting, could be initiated over hundreds of thousands of hectares. Houghton (1990) estimated that about 500 million ha of degraded lands might be reforested and another 365 million ha of fallow allowed to return to forests if shifting cultivation were successfully replaced with low-input permanent agriculture (Sanchez and Benites 1987). If agroforestry were implemented on large areas of croplands and pastures, the area available for accumulation of carbon in woody biomass would be approximately 1.9 billion ha (Houghton, Unruh, and Lefebvre 1993), although the accumulation of carbon per unit area would be considerably less than it is in forests. Perhaps as much as 1.5 BMT carbon could be accumulated annually in woody biomass if huge areas of land were to be replanted. This strategy would reduce total emissions of carbon to about 6.5 BMT. Requiring an enormous amount of land, energy, and ex-

Table 15.5 Potential annual emissions of carbon to or from the atmosphere from human activities

Strategy	Fossil fuels	Deforestation	Reforestation[a]	Sustainable use harvest of fuel	Total emissions	Potential reduction from present
I	6.0	2.0	<0.1	0	8.0	0%
II	6.0	2.0	−1.5	0	6.5	20%
III	6.0	0.0	<0.1	0	6.0	25%
IV	0.0	2.0	0.0	0	0.0	75%

Note: All values in BMT/yr. Positive values indicate emissions; negative values indicate a removal of carbon from the atmosphere.

I: This strategy represents 1990 emissions of carbon.

II: Reforestation: 150 BMT carbon might be stored in new plantations, in forests protected from further logging and shifting cultivation, and in agroforestry. The accumulation of carbon is assumed arbitrarily to take place over 100 years.

III: No deforestation.

IV: Replacing fossil fuels with wood-based fuels grown sustainably (assuming that world energy consumption will not increase substantially above 1990 rates). Combustion of wood fuels emits as much, or more, carbon to the atmosphere as combustion of fossil fuels. The emissions from wood, however, are balanced by accumulations of carbon in the forests growing to provide future fuel.

[a] This annual accumulation of carbon in growing forests persists only while the forests are growing or while new lands are being reforested. Once forests have regrown (in 15 to 100 years) they continue to hold carbon, but they no longer withdraw it from the atmosphere.

pense, such reforestation would accumulate carbon only while the forests were growing. Once grown, the forests would hold, perhaps, 150 BMT carbon but would no longer withdraw carbon from the atmosphere. Reforestation would stabilize atmospheric concentrations of CO_2 only temporarily, at best.

A third strategy would completely halt deforestation. Total emissions would thereby be reduced by about 2 BMT per year. The second and third strategies together would almost halve the annual emissions of carbon to the atmosphere.

The fourth strategy, the most effective way to use forests to manage atmospheric carbon, would substitute sustainably grown wood fuels for fossil fuels worldwide. Burning wood does not eliminate emissions, but under sustainable management the carbon emitted from burning wood is accumulated by the forests growing next year's fuel. Wood would not be used as the end fuel; it would be converted to ethanol, hydrogen, or some other fuel suitable for transport and efficient use. This strategy would eliminate the major contributor of

CO_2 emissions. Total emissions would be only 2 BMT carbon annually if deforestation continued. If the substitution of wood for fossil energy were combined with a halt to deforestation, the global emissions of carbon could be reduced, theoretically, to zero. Increased efficiency of energy use and use of other renewable energy sources would, of course, reduce the requirements for wood-derived fuels.

The intent of presenting these strategies is to show the large amounts of carbon and land, and the magnitude of the changes, involved in stabilizing concentrations of CO_2 in the atmosphere. Industrial emissions of greenhouse gases must be reduced, but so must other sorts of emissions. Although the industrialized countries may have caused the problem, the problem itself is likely to affect countries and peoples everywhere.

IPCC CLIMATE PROJECTIONS

The IPCC's business-as-usual scenario (Houghton, Jenkins, and Ephraums 1990), assuming that current growth rates of emissions continue, predicts that the mean global temperature will increase by about 0.3°C per decade in the twenty-first century (the range of uncertainty includes 0.2°C to 0.5°C per decade). The average warming will be about 1°C by 2025 and 3°C by 2100. Sea level is predicted to rise about 6 cm per decade (with a range of uncertainty of 3 to 10 cm per decade), increasing about 65 cm by 2100. If emissions are controlled, the predicted increase in global mean temperatures is 0.1°C to 0.2°C per decade and the predicted rise in sea level is also reduced (Houghton, Jenkins, and Ephraums 1990).

These global averages will not apply in individual regions. Mid-latitude regions may warm more than the tropics, but predictions vary among models. Regional changes in precipitation and soil moisture, probably more important than temperature for most of the tropics, are even less consistently predicted. Thus changes in climate for particular regions are uncertain.

These sorts of predictions are not very compelling except to low-lying coastal communities. But the fact that scientists cannot yet make specific predictions does not imply that the changes will be minor. The possible effects include massive disturbance of climate, with associated disruption of agricultural production and political stability. The world's population is expected to double in forty or fifty years. Is there enough arable land to provide twice the food now produced? The answer could be yes if new lands brought into agriculture were managed sustainably, but the increasing area of degraded lands in the tropics shows this not to be the case (Houghton 1994). Will new strains of crops help double production? Will the effects of the warming together with increased levels of atmospheric carbon dioxide actually increase agricultural production, as some scientists predict? We can hope for these types of relief. But hope is a poor strategy when the risks are unacceptable. Add to the growing world population a changing climate with year-to-year variation in rainfall, frosts, droughts, and other unpredictable events, and the picture is not reassuring.

The following aspects of climatic change, though not quantitative, are cause for concern:

1. The warming is likely to be more *rapid* than ever experienced in human history, unless major steps are taken now to reduce emissions of greenhouse gases. Rapid change is difficult to predict, and rapid environmental alterations are difficult to prepare for and adjust to.

2. The warming will be *continuous*. Any policy of dealing with climatic change that is based on adaptation or coping will find itself several steps behind the current climate. Rapid and continuous change together mean that adjustments will be made to environments that are already past; the current climate will be continually new.

3. The changes will be *irreversible* within a human lifetime. If the change turns out to be unacceptable, there will be no way to undo it for decades. If all releases of greenhouse gases were stopped tomorrow, for example, the earth would continue to warm another 0.5°C or more. The longer greenhouse gases are emitted to the atmosphere, the larger this commitment to a further temperature rise will become. Policies proposed to reduce climatic change address only future emissions of greenhouse gases. Nothing but time will be able to remove gases that have already been released to the atmosphere.

4. Finally, the change is almost *open ended*. Carbon dioxide concentrations will not stop at a doubling, as experiments with global climate models suggest, unless deliberate policies are enacted to reduce further emissions. Once burned, recoverable reserves contain enough fossil fuel to raise the atmospheric concentrations of carbon dioxide by a factor of 5 to 10 above preindustrial levels. If all forests are cleared, the trees and soils of the earth will release enough carbon to double or triple atmospheric concentrations.

LOCAL AND REGIONAL EFFECTS OF DEFORESTATION

All the discussion linking deforestation to changes in climate has, thus far, considered only the global effects of increased emissions of greenhouse gases. Evidence from both direct observation (see reviews by Myers [1988] and Clark [1992]) and simulations with general circulation models (GCMs) (Dickinson and Henderson-Sellers 1988; Lean and Warrilow 1989; Mylne and Rowntree 1992;

Nobre, Sellers, and Shukla 1991; Shukla, Nobre, and Sellers 1990) suggests that deforestation also affects local climates, not through the emission of greenhouse gases but through direct modification of energy and water budgets.

The replacement of tropical forests with grasslands increases local air and soil temperature, decreases evapotranspiration (the water released to the atmosphere through vegetation and surface evaporation), and decreases precipitation (Lean and Warrilow 1989; Nobre, Sellers, and Shukla 1991; Shukla, Nobre, and Sellers 1990). Excess water is carried away by streams and rivers and added to the groundwater. Evapotranspiration is reduced, in part because less radiant energy is absorbed by grasslands than by forests, in part because atmospheric turbulence (surface roughness) is greater above forests than above grasslands and hence can remove water more rapidly from forests, and in part because the roots of trees generally penetrate to deeper layers of soil than the roots of pastures and hence have access to more water (Nepstad et al. 1995). If less water is available, the vegetation will respond by closing stomata, thereby increasing the resistance to evapotranspiration and increasing temperatures. Temperature is increased after the removal of forest because evapotranspiration, a cooling process, is reduced. The warming that accompanies reduced evapotranspiration more than compensates for the cooling expected from the increased albedo (proportion of radiant energy reflected) that accompanies replacement of forest with grassland or soil.

The decrease in evapotranspiration as a result of deforestation has further consequences for local and regional climate. Evapotranspiration not only cools the surface but also adds moisture to the atmosphere. If the supply of moisture is reduced, precipitation is also reduced either locally or downwind. In the Amazon basin, for example, 50–75% of the precipitation is recycled within the basin (Salati and Vose 1984). The reduction in precipitation, in turn, reduces evapotranspiration further and increases temperature further because less water is available for the plants. Less evaporation means fewer clouds and more sunshine. Increased temperatures result not only from a further reduction in evapotranspiration but from an increase in the hours of sunshine. Clark (1992) estimates that almost 40% of the increase in annual temperature associated with deforestation is from an increase in hours of direct sunlight. The other 60% is from reduced evapotranspiration.

When tropical forests of the Amazon were replaced with degraded grasslands (pastures) in a model simulation, mean surface temperature increased by about 2.5°C, evapotranspiration decreased by about 30%, and precipitation decreased by about 25% (Nobre, Sellers, and Shukla 1991). The changes were larger in the dry season, which was lengthened in the southern part of the Amazon. These changes are generally larger than those predicted from an enhanced greenhouse effect. Nobre, Sellers, and Shukla warned that a complete and rapid deforestation of Amazonia could irreversibly change the southern part of the region. Reductions in rainfall could be so large that forests, once removed, might not be reestablished.

Nobre, Sellers, and Shukla (1991) wondered whether such irreversible changes might also occur in tropical Africa or Southeast Asia. In Africa the elimination of forests at the northern and southern edges, whether induced by human activity or by larger-scale changes in climate, might lead to a further warming and drying in these locations and to subsequent loss of forests. In Southeast Asia the feedback among forests, land use, and climate is less clear. The monsoon rains are driven by large-scale interactions between the oceans and the atmosphere, so reductions in the area of forests might not affect the rains noticeably. Rates of heating between the land surface and the ocean surface are also important in controlling the monsoons, however, and deforestation might well change these rates. Possible links between deforestation and monsoon climates are not well understood, although Myers (1988) gives examples of local climatic changes attributed to deforestation in India, Malaysia, and the Philippines.

Deforestation may be even more deleterious locally and nationally than it is globally. Preservation of forested land helps not only to reduce the rate at which carbon accumulates in the atmosphere but also to reduce regional environmental variability.

CONCLUSIONS

Forests are converted to agricultural land to provide food and revenue for expanding numbers of people. At present rates, the supply of new agricultural land from this source will last less than a century in tropical regions. Forests are also used to provide local and national revenue through timber and fuel production. If forests are managed sustainably, revenues from these products may continue indefinitely. The large reservoir of biodiversity in tropical forests is potentially beneficial both locally and worldwide, but these products are largely unvalued. Forests also slow the rate of increase of heat-trapping gases in the atmosphere and regulate climate at local, regional, and global levels. As regulators of climate, they affect not only rainfall and temperature but such local processes as runoff, groundwater recharge, siltation, and flooding. These processes clearly affect the entire human enterprise, yet the stabilization of soil and of water flows, for example, is not often counted in the valuation of forests. Instead, forests are valued largely for their products.

In addition to goods and services (Ramakrishna and Woodwell 1993), forests may provide a variety of functions that scientists know little about, concerning the chemical and physical stability of the earth's atmosphere. In the face of this ignorance, the best policy is perhaps a precautionary one. Some fraction of each continent's land surface should remain forested. The fraction is difficult to specify but may depend on stability of climate desired by the earth's inhabitants. In the meantime, before scientists can define the requirements for climate stability, before society learns the value of forests' goods and services, and before short-term interests eliminate large areas of the remaining forests, some combination of national and global interests must agree on an insurance policy that preserves forests.

The international conventions on climate and biodiversity recognize a public interest that transcends local and national interests. These conventions will address forests, but their examination may be limited to forests' roles in producing nonfossil fuels, for example, or storing carbon. Other known and unknown functions of forests may be overlooked. Eventually, an international convention on forests will be needed (Ramakrishna and Woodwell 1991). First, however, industrial countries must not turn to poorer countries for solutions to problems that are largely of the industrial world's making, and developing countries must recognize that their paths and definitions of development are not consistent with sustainable use of natural resources. The world we want is considerably different from the one we are creating.

REFERENCES

Brown, S., A. J. R. Gillespie, and A. E. Lugo. 1989. Biomass estimation methods for tropical forests with applications to forest inventory data. *For. Sci.* 35: 881–902.
———. 1991. Biomass of tropical forests of South and Southeast Asia. *Can. J. For. Res.* 21: 111–117.
Brown, S., and A. E. Lugo. 1982. The storage and production of organic matter in tropical forests and their role in the global carbon cycle. *Biotropica* 14, 3: 161–187.
Bruenig, E. F. 1974. *Ecological Studies in the Kerangas Forests of Sarawak and Brunei.* Borneo Literature Bureau, Sarawak, Malaysia.
Clark, C. 1992. Empirical evidence for the effect of tropical deforestation on climatic change. *Env. Cons.* 19: 39–47.
Darmstadter, J. 1971. *Energy in the World Economy: A Statistical Review of Trends in Output, Trade and Consumption since 1925.* The Johns Hopkins University Press, Baltimore.
Detwiler, R. P. 1986. Land use change and the global carbon cycle: The role of tropical soils. *Biogeochem.* 2: 67–93.
Dickinson, R. E., and A. Henderson-Sellers. 1988. Modelling tropical deforestation: A study of GCM land-surface parametrizations. *Q. J. Roy. Met. Soc.* 114: 439–462.

Edwards, P. J., and P. J. Grubb. 1977. Studies of mineral cycling in a montane rain forest in New Guinea, I: The distribution of organic matter in the vegetation and soil. *J. Ecol.* 65: 943–969.

FAO (Food and Agriculture Organization). 1973. Forestry and forest industries development. Malaysia: A national forest inventory of West Malaysia 1970–1972. FO:DP/MAL/72/009. Tech. Rep. No. 5., FAO/UNDP, Rome.

———. 1993. Forest resources assessment 1990. Tropical countries. FAO Forestry Paper No. 112, FAO, Rome.

FAO/UNEP. 1981. Tropical forest resources assessment project. FAO, Rome.

Fearnside, P. M., A. T. Tardin, and L. G. M. Filho. 1990. *Deforestation Rate in Brazilian Amazonia.* National Secretariat of Science and Technology, Brasília, Brazil.

Flint, E. P., and J. F. Richards. 1994. Trends in carbon content of vegetation in South and Southeast Asia associated with changes in land use. In *Effects of Land Use Change on Atmospheric CO₂ Concentrations: South and Southeast Asia as a Case Study.* V. H. Dale, editor. Springer-Verlag, New York.

Goldemberg, J., T. B. Johansson, A. K. N. Reddy, and R. H. Williams. 1985. An end-use oriented global energy strategy. *Ann. Rev. Energ.* 10: 613–688.

Government of Malaysia. 1987. *Inventori hutan Nasional II Semenanjung Malaysia 1981–1982.* [Second national forest inventory of Peninsula Malaysia 1981–1982]. Unit Pengurusan Hutan, Ibu Perhutan. Semanenanjung Malaysia, Kuala Lumpur.

Hall, C. A. S., and J. Uhlig. 1991. Refining estimates of carbon released from tropical land-use change. *Can. J. For. Res.* 21: 118–131.

Houghton, J. T., B. A. Callander, and S. K. Varney, editors. 1992. *Climate Change 1992: The Supplementary Report to the IPCC Scientific Assessment.* Cambridge University Press, Cambridge.

Houghton, J. T., G. J. Jenkins, and J. J. Ephraums, editors. 1990. *Climate Change: The IPCC Scientific Assessment.* Cambridge University Press, Cambridge.

Houghton, R. A. 1990. The future role of tropical forests in affecting the carbon dioxide concentration of the atmosphere. *Ambio* 19: 204–209.

———. 1991. Tropical deforestation and atmospheric carbon dioxide. *Climatic Change* 19: 99–118.

———. 1994. The worldwide extent of land-use change. *BioSci.* 44: 305–313.

Houghton, R. A., R. D. Boone, J. R. Fruci, J. E. Hobbie, J. M. Melillo, C. A. Palm, B. J. Peterson, G. R. Shaver, G. M. Woodwell, B. Moore, D. L. Skole, and N. Myers. 1987. The flux of carbon from terrestrial ecosystems to the atmosphere in 1980 due to changes in land use: Geographic distribution of the global flux. *Tellus* 39B: 122–139.

Houghton, R. A., and J. L. Hackler. 1994. The net flux of carbon from deforestation and degradation in South and Southeast Asia. In *Effects of Land Use Change on Atmospheric CO₂ Concentrations: South and Southeast Asia as a Case Study.* V. H. Dale, editor. Springer-Verlag, New York.

Houghton, R. A., D. L. Skole, and D. S. Lefkowitz. 1991. Changes in the landscape of Latin America between 1850 and 1980, II: A net release of CO_2 to the atmosphere. *For. Ecol. & Mgmt.* 38: 173–199.

Houghton, R. A., J. D. Unruh, and P. A. Lefebvre. 1993. Current land use in the tropics and its potential for sequestering carbon. *Glo. Biogeochem. Cyc.* 7: 305–320.

Hozumi, K., K. Yoda, and T. Kira. 1969. Production ecology of a tropical rainforest in southwestern Cambodia, II: Photosynthetic production in an evergreen seasonal forest. *Nat. Life Southeast Asia* 6: 57–81.

Kira, T. 1978. Community architecture and organic matter dynamics in tropical lowland rain forests with special reference to Pasoh Forest, West Malaysia. In *Tropical Trees as Living Systems.* P. B. Tomlinson and M. H. Zimmermann, editors. Cambridge University Press, Cambridge.

Lean, J., and D. A. Warrilow. 1989. Simulation of the regional climatic impact of Amazon deforestation. *Nature* 342: 411–413.

Meher-Homji, V. M. 1991. Probable impact of deforestation on hydrological processes. *Climatic Change* 19: 163–173.

Myers, N. 1980. *Conversion of Tropical Moist Forests*. National Academy of Sciences Press, Washington, D.C.

———. 1988. Tropical deforestation and climatic change. *Env. Cons.* 15: 293–298.

———. 1991. Tropical forests: Present status and future outlook. *Climatic Change* 19: 3–32.

Mylne, M. F., and P. R. Rowntree. 1992. Modelling the effects of albedo change associated with tropical deforestation. *Climatic Change* 21: 317–334.

Nepstad, D. C., C. R. de Carvalho, E. A. Davidson, P. H. Jipp, P. A. Lefebvre, G. H. Negreiros, E. D. da Silva, T. A. Stone, S. E. Trumbore, and S. Vieira. 1995. The role of deep roots in the hydrological and carbon cycles of Amazonian forests and pastures. *Nature* 372: 666–669.

Nobre, C. A., P. J. Sellers, and J. Shukla. 1991. Amazonian deforestation and regional climate change. *J. Clim.* 4: 957–988.

Ogawa, H., K. Yoda, T. A. Kira, K. Ogino, T. Shidei, D. Ratanawongse, and C. Apasuty. 1965. Comparative ecological studies on three main types of forest vegetation in Thailand, I: Structure and floristic composition. *Nat. Life Southeast Asia* 4: 13–48.

Palm, C. A., R. A. Houghton, J. M. Melillo, and D. L. Skole. 1986. Atmospheric carbon dioxide from deforestation in Southeast Asia. *Biotropica* 18: 177–188.

Post, W. M., W. R. Emanuel, P. J. Zinke, and A. G. Stangenberger. 1982. Soil carbon pools and world life zones. *Nature* 298: 156–159.

Proctor, J., J. M. Anderson, P. Chai, and H. Vallack. 1983. Ecological studies in four contrasting lowland rain forests in Gunung Mulu National Park, Sarawak. I. Forest environment, structure and floristics. *J. Ecol.* 71: 237–260.

Ramakrishna, K., and G. M. Woodwell. 1991. *Report of the Workshop on the Conservation and Utilization of World Forests*. Woods Hole Research Center, Woods Hole, Mass.

Ramakrishna, K., and G. M. Woodwell, editors. 1993. *World Forests for the Future: Their Use and Conservation*. Yale University Press, New Haven.

Richards, J. F., and E. P. Flint. 1994. A century of land use change in South and Southeast Asia. In *Effects of Land Use Change on Atmospheric CO_2 Concentrations: South and Southeast Asia as a Case Study*. V. H. Dale, editor. Springer-Verlag, New York.

Rotty, R. M., and G. Marland. 1986. Fossil fuel combustion: Recent amounts, patterns, and trends of CO_2. In *The Changing Carbon Cycle: A Global Analysis*. J. R. Trabalka and D. E. Reichle, editors. Springer-Verlag, New York.

Sabhasri, S. 1978. Effects of forest fallow cultivation on production and soil. In *Farmers in the Forest: Economic Development and Marginal Agriculture in Northern Thailand*. P. Kunstadter, E. C. Chapman, and S. Sabhasri, editors. University Press of Hawaii, Honolulu.

Salati, E., and P. B. Vose. 1984. Amazon basin: A system in equilibrium. *Science* 225: 129–138.

Sanchez, P. A., and J. R. Benites. 1987. Low-input cropping for acid soils of the humid tropics. *Science* 238: 1521–1527.

Schlesinger, W. H. 1977. Carbon balance in terrestrial detritus. *Ann. Rev. Ecol. Syst.* 8: 51–81.

———. 1986. Changes in soil carbon storage and associated properties with disturbance and recovery. In *The Changing Carbon Cycle: A Global Analysis*. J. R. Trabalka and D. E. Reichle, editors. Springer-Verlag, New York.

Setzer, A. W., and M. C. Pereira. 1991. Amazonia biomass burnings in 1987 and an estimate of their tropospheric emissions. *Ambio* 20: 19–22.

Shukla, J., C. Nobre, and P. Sellers. 1990. Amazon deforestation and climate change. *Science* 247: 1322–1325.

Skole, D., and C. Tucker. 1993. Tropical deforestation and habitat fragmentation in the Amazon: Satellite data from 1978 to 1988. *Science* 260: 1905–1919.

Yoda, K., and T. Kira. 1969. Comparative ecological studies on three main types of forest vegetation in Thailand, V: Accumulation and turnover of soil organic matter, with notes on the altitudinal soil sequence on Khao (Mt.) Luang, peninsular Thailand. *Nat. Life Southeast Asia* 6: 57–81.

Zinke, P. J., S. Sabhasri, and P. Kunstadter. 1978. Soil fertility of the Lua forest fallow system of shifting cultivation. In *Farmers in the Forest: Economic Development and Marginal Agriculture in Northern Thailand*. P. Kunstadter, E. Chapman, and S. Sabhasri, editors. University Press of Hawaii, Honolulu.

CONTRIBUTORS

Abang Haji Kassim bin Abang Morshidi, Sarawak Forest Department, Wisma Sumber Alam, Jalan Stadium, 93660 Kuching, Sarawak, Malaysia

Jo Anderson, Department of Biological Sciences, Exeter University, Exeter EX4 4QG, England

Elizabeth L. Bennett, Wildlife Conservation International, 7 Jalan Ridgeway, 93200 Kuching, Sarawak, Malaysia

Eberhard F. Bruenig, Research Section, Sarawak Forest Department, Jalan Stadium, Wisma Sumber Alam, 93660 Kuching, Sarawak, Malaysia

Hamid Bugo, Government of Sarawak, 93760 Kuching, Sarawak, Malaysia

Zainuddin Dahaban, Wildlife Conservation International, 7 Jalan Ridgeway, 93200 Kuching, Sarawak, Malaysia

Stephen D. Davis, Centres of Plant Diversity Project, Royal Botanic Gardens, Kew, Richmond, Surrey TW9 3AE, England

Eric Dinerstein, Conservation Science Program, World Wildlife Fund, 1250 24th Street, NW, Washington, DC 20037–1175

Hans J. Droste, World Forestry, Hamburg University, Leuschnerstrasse 91, 21031 Hamburg, Germany

Mark Forney, Conservation Science Program, World Wildlife Fund, 1250 24th Street, NW, Washington, DC 20037–1175

Melvin Terry Gumal, Sarawak Forest Department, Wisma Sumber Alam, Jalan Stadium, 93660 Kuching, Sarawak, Malaysia

R. A. Houghton, Woods Hole Research Center, P. O. Box 296, Woods Hole, MA 02543

Kuswata Kartawinata, Regional Office for Science and Technology of Southeast Asia, UNESCO, Jalan M. H. Thamrin 14, Tromolpos 1273, Jakarta, Indonesia

Thomas E. Lovejoy, Smithsonian Institution, Washington, DC 20560

Pedro H. Moura-Costa, Rakyat Berjaya–Face Foundation, Rainforest Rehabilitation Project, Innoprise Corporation Sdn Bld., Danum Valley Field Center, PS 282, 91108 Lahad Datu, Sabah, Malaysia

Ruth Nussbaum, Danum Valley Field Centre, PS 282, 91108 Lahad Datu, Sabah, Malaysia

Junaidi Payne, WWF-Malaysia, WDT No. 40, 89400 Likas, Sabah, Malaysia

Richard B. Primack, Department of Biology, Boston University, 5 Cummington Street, Boston, MA 02115

John Proctor, Department of Biological and Molecular Sciences, University of Stirling, Stirling FK9 4LA, Scotland

Herwasono Soedjito, Research and Development Center in Biology, Indonesian Institute of Sciences, Jl. Ir. H. Juanda 22, Bogor 16122, Indonesia

E. Soepadmo, Forest Research Institute of Malaysia, Kepong, 52109 Kuala Lumpur, Malaysia

Tom Spencer, Department of Geography, Cambridge University, Cambridge CB2 3EN, England

Jeffrey R. Vincent, Harvard Institute for International Development, Harvard University, Cambridge, MA 02138

T. C. Whitmore, Department of Geography, University of Cambridge, Downing Place, Cambridge CB2 3EN, England

Eric D. Wikramanayake, Department of Herpetology, National Zoological Park, Washington, DC 20008

INDEX

Page numbers in italics refer to illustrations

African tropical rainforests: climate of, 5–7, 285; distribution of, 5–10; floristic composition of, 8–9; forest formations in, 9–10; and hardwood trade, 12–14; land-use changes in, 277–278; map of, *6;* species richness of, 7–8

Agriculture: and climate disturbance, 282; commercial, 235; conversion of forests to, 245, 250, 263, 285; effects on wildlife, 66; and forest production, 206, 280; low-productivity subsistence farming, 226; potential of, 237–238; slash-and-burn, 48, 105, 182; threat of, 208; and Totally Protected Areas (TPAs), 212; in tropical moist forests, 140; and vertebrate distribution, 60

Alan bunga forests, 43–44

Alan forests, 42–43, 45, 48

Altitude, and vertebrate distribution, 61–62

Amazon rainforests, 7–8, 13–14, 42, 46, 276–277, 284–285

American tropical rainforests: climate of, 5–7, 284–285; deforestation of, 276–278; distribution of, 5–10; floristic composition of, 8–9; forest formations in, 9–10; and hardwood trade, 13–14; map of, *6;* seed dispersal in, 59; soils of, 92–93; species richness of, 7–8; and topography, 91; vegetation types of, 89

Amphibians, 54, 142

Anacardiaceae, 28

Animal-plant interactions, 54, 59–61, 69–70, 83–84, 133

Annonaceae, 8, 20, 27, 59

Apical dominance, 122–123

Apo Kayan-Mentarang Biosphere Reserve, 134

Apocynaceae, 20, 27

Aquaculture, 212

Aquatic ecology, *136*

Aquatic plants, 20

Araceae, 27

Asclepiadaceae, 20

Asian rhinoceroses, 58

Australia, 95–96, 142–144, 158, 166, 269

Bakau, 55

Bako National Park, *90,* 93, *132,* 205–206

Bali, 145–146, 157, 166

Banana, 55

Bangladesh, 10, 157

Banteng, 82

Barito Ulu, 94, 133

Barking deer, *69,* 77

Barro Colorado Island, 131–132

Batang Ai National Park, 32, 67–*68,* 72, 79, 207, 211–212

Bats, 54–*56,* 60, 63

Batu Laga Wildlife Sanctuary, 211

Belum forest, 31

Bindang, 9, 45, 49

Bintangor, 9, 11, *46*

Bioassay experiments, 94–95

Biogeographic units: and Conservation Potential/Threat Index (CPTI), 148, 166–167;

Biogeographic units *(continued)*
    lowland rainforest distribution in, 150–152, 171–172; patterns of, 149–151; protected areas of, 151–156; tropical moist forest distribution in, 149–151, 166–167, 171–172

Biological diversity: assessment of, 142–143; conservation of, 135, 140, 142–143, 159–161, 165–168, 178; context of, 135; decline of, 140; defined, 205; and human needs, 135; of Indonesia, 129; investment priorities for, 143, 158–162, 165–166; and logging, 247, 259; and long-term research, 134; national strategies for, 167–168; and natural resources demands, 184; needs for data on, 167, 209; and process, 135; in Sarawak, 204–218; threats to, 141; and Totally Protected Areas (TPAs), 207–208; and tropical moist forests, 142–143, 158, 162. *See also* Plant diversity; Species richness

Biomass: and carbon, 34, 273; estimates of, 87; and logging, 111; in Malesian tropical rainforests, 34; nutrients in, 87–88; reductions in, 274–275; and soils, 92; and sustainability, 46–48

Biotic carbon, 269

Birds: conservation of, 135; endemic species of, 149–150, 177; and fig plants, 59; hunting of, 76–77, 82; migrant, 209; nesting sites of, 60; on shifting-cultivation land, 72; and pollination, 54; and rainforest ecosystems, 54; sanctuaries for, 31; as seed dispersers, 58; species richness of, 142, 148–150; and timber production, 62; and tree plantations, 62–63. *See also names of specific birds*

Bombacaceae, 28, 59

Boom-and-bust forest sector development, 241–244, 259

Borneo: birds of, 58; dipterocarp trees in, 11; endemic species of, 23; and fire, 63; forest formations in, 9; mammals of, 54, 61; plant diversity in, 19, 21; protected areas of, 153, 158, 166, 208, 210

Branch parasites, 20

Brazil, 106, 171, 177, 276–277

Breeding populations, 61–64, 81–82

Brunei, 21–22, 30, 91, 144, 155, 158, 166

Bukit Baka Bukit Raya National Park, 133

Bukit Mersing, 93

Bulbuls, 58, 63, 75, 81

Burial grounds, 233–234

Burma, 28, 143–145, 165–166

Burmese brown tortoise, 75

Burseraceae, 20, 59

Butterflies, 142

Caesalpinaceae, 8

Cambodia, 144, 155, 158, 162

Canopy: characteristics of, 42–44; and debris piles, 106; gaps in, 11–12, 44, 93, 110–111, 122, 224; height of, 45; plants within, 20; preservation of, 49; role of, 34, 50

Captive breeding, 63–64

Carbon: and biomass, 34, 273; biotic, 269; emissions of, 268–269, 271–278, 281; and land-use changes, 269–271, 274; management of, 280–282

Carbon dioxide, 265–266, 268–269, 278–283

Carbon monoxide, 278

Cats, 78, 83

Celebes, 23

Central Indochina, 153

Centres of Plant Diversity (CPD) project: site analysis, 181–184; site identification, 177–178, 185; site selection, 179–181; and threats to plants, 176–177; tropical Asian sites, 186–202. *See also* Plant diversity

Chempedak, *24*

China, 28, 142–143, 151, 157, 165–166, 181

Chlorofluorocarbons, 265–266, 278–279

Chrysobalanaceae, 100

Civets, 57, 78, 83

Clearing methods, 106

Climate: and agriculture, 282; change in, 263–269; and ecosystems, 33–34; global warming, 34, 265–267, 278–279; local, 284; in Malesian tropical rainforests, 19–20, 33–34; projections of, 282–283; and soils, 89; and tropical rainforests, 5–7, *6*, 263–264, 285–286; and warming, 265–266

Climbing bamboos, 110–112

Clonal propagation, 121–123

Closed soil systems, 88

Clouded leopard, 70, 209

Clusiaceae, 20, 28

Colonizer species, 81

Comparative advantage theory, 257–258

Conifers, 9, 19

Connaraceae, 20

Conservation: aims of, 81; and biogeographic units, 148, 166–167; of biological diversity, 135, 140, 142–143, 159–161, 165–168, 178; of dipterocarp forests, 32; and economics, 137; education, 137; ex situ, 208; financing of, 162, 165; and human resources, 134, 182–184, 209; implementation of, 178–179; in situ, 205; in Indonesia, 133, 162, 167; investments of, 143, 159–161, 165–166; and limestone forests, 98; of Malesian tropical rainforests, 29–36; of mammals, 135, 210; of plant diversity, 29–36, 176–202; political/social dimensions of, 178, 184, 212; priorities of, 141, 178; and protected areas, 30; and rainforest soils, 87–100; recommendations for, 162–166; regional patterns of, 143–147; site analysis, 181–184; site selection, 179–181; of Southeast Asian tropical rainforests, 54–64; status of, 182–184; and timber harvesting, 223–224; of tropical moist forests, 140–141, 159–168; and vertebrates, 30, 62–64, 135

Conservation Potential/Threat Index (CPTI): and biogeographic units, 148, 166–167; and biological diversity, 142–143, 178; and investment priorities, 140; and landscape analyses, 143, 155–158; and protected areas, 147; recommendations from, 162–166; and tropical moist forests, 141–143, 147

Context, of biological diversity, 135

Corocovado National Park, 21

Costa Rica, 252

CPD. *See* Centres of Plant Diversity (CPD) project

CPTI. *See* Conservation Potential/Threat Index

Crocker Range, 32

Customary rights, 165, 212, 234, 238

Cutting limits, 48, 69, 105, 223, 255–257

Cuttings (propagation), 118–120, *119*, 123

Cuyabeno, 8

Danum Valley, 32, 55, 87, 117–119, *118*

Debris piles, and logging, 106–107, 109

Deer, 57, *69,* 77, 80

Deforestation: and biomass reduction, 275; and carbon, 269; and carbon emissions, 272–275, 281; and conservation, 178; and fragmentation of forests, 157; and global warming, 278–279; and greenhouse effect, 279; impact of, 143; local effects of, 283–285; rates of, 141–142, 250, 272–278

Degraded soils, 110

Deposition sites, 108

Dermakot Forest Reserve, 111

Diameter at breast height (dbh), 69, 105, 223

Dioscoreaceae, 27

Dipterocarp forests: conservation of, 32; ecological studies of, 135; enrichment planting of, 116–123; and fertilization experiments, 95; floristic variation within, 91; logging of, 80, *241;* and Malayan uniform system, 250; seedlings of, 105–113, 122; soils of, 89; in Southeast Asian rainforests, 9–12, 14, 20; studies of, 67–70; sustainable management of, 49; timber harvesting in, *68,* 223; and vertebrate distribution, 61; wood production of, 93

Dipterocarp trees: clonal propagation of, 121–122; extinction of, 25; fruiting of, 11–12, 117; genetic improvement of, 121–122; growth rates of, 123; in Indonesia, 130; micropropagation of, 120–121; in Sarawak, 208; seeds of, 117–118; and soils, 99; as timber, 11; vegetative propagation of, *118*–120; wildings of, 117–118

Disturbed forest, 105–107

Dominant-use management, 258–259

Doves, 58

Droughts, 116, 123, 264

Dugongs, 207

Durian, 55

Dyes, 55, 176

Ebenaceae, 8, 100

Ecological studies: and human resources, 134; in Indonesia, 130–134, 137; limitations of, 131–133; long-term, 131–137; of Malesian tropical rainforests, 20–22; of Sarawak, *132*

Economics: boom-and-bust pattern,

Economics *(continued)*
242–244, 259; and conservation, 137; and ecosystems, 221; and forest management, 253–258; and forest sector development, 241–242; and fruit trees, 28, 33, 55; and medicinal plants, 27, 33, 55; and ornamental plants, 27–28, 33; and rattans, 28–29, 33; and second-growth forest management, 253–258; and timber depletion, 244–247; and timber industry, 227, 229–230, 237, 241–259; and timber production, 247–253, 259; and timber trees, 25–27, 33, 55

Ecosystems: and climate, 33–34; and economic development, 221; and fertilization, 95; and habitats, 208; and hunting, 82–83; in situ conservation of, 205; and logging, 48, 62; mangrove, 131; and plant diversity, 19–20; plant-animal interactions, 59–60, 83–84; plants as basis of, 176; services of, 263; and vertebrates, 54–64, 82

Ecotourism, 168. *See also* Tourism

Ecuador, 177

Edge species, 81

Education, environmental, 207, 212

Elephants, 58–59, 61

Employment, 206, 230–231

Endangered species, 207, 209, 211–212

Endemic species: assessment of, 142; conservation of, 177–181; in Indo-Pacific region, 142, 148–155, 165; in Malesian tropical rainforests, 21–25, 32; in Sarawak, 208; and Totally Protected Areas (TPAs), 208–209

Enrichment planting: of dipterocarps, 116–123; ecological considerations, 123; and forest regeneration, 113, 253; sources of planting material, 117–122

Environmental education, 207, 212

Epiphytes, 20

Erosion. *See* Soil erosion

Euphorbiaceae, 20, 27, 59, 100

Evapotranspiration, 284

Even-flow sustainability, 250–251, 259

Evergreen Caatinga forest, 42

Ex situ conservation, 208

Export prices, 244

Ferns, 19, 22, 83

Fertilization, 94–96, 112–113

Fibers, 55, 176

Fig plants, 55, 59

Fiji, 143, 148, 151

Fires, 63, 98–99

Fishes, 54, 59, 142

Floods, 264

Floristic composition, 8–9

Flowering plants, 19–20, 22

Flowerpeckers, 54

Flying foxes, 82

Flying squirrels, 60

Foliar analyses, 93

Food supply, 59–60, 62, 226

Forage crops, 176

Forest dynamics, 129, 133

Forest management: and carbon loss, 273; dominant-use, 258–259; and ecological research, 131–137; economically viable, 253–258; funds for, 253; and human resources, 134; multiple-use management, 258; natural, 50; of protected areas, 158, 162, 165, 182–184; in Sarawak, 221–224, 239, *239;* and scarcity, 245; of second-growth forests, 253–258; and soils, 50–51, 87–100; and sustainability, 48–50, 129–138, 209, 222–223, 244; technical, 133; and timber trade, 241–259; wildlife responses to, 63–64, 66–84. *See also* Forest regeneration; Timber industry

Forest products, 25–29, 33, 54–55, 116, 206. *See also* Timber industry

Forest regeneration: and animal-plant interactions, 84, 133; and enrichment planting, 113, 253; and logging, 12; management of, 116; and soils, 109–112; and sustainability, 248. *See also* Deforestation; Forest management

Forest resources, 221–222, 227, 229–230, 237. *See also* Timber industry

Forest sector. *See* Timber industry

Forest stature, 92–93

Forests. *See* Southeast Asian tropical rainforests; Tropical rainforests; *other specific types of forests*

Fossil fuels, 265, 268, 271, 281–282

Fourier series analysis, 70

Freshwater forests, 31–32
Fruit bats, 54–56, *56*
Fruit trees, 28, 33, 55, *210*, 234
Fuelwood, 55, 176, 271, 281–282

Gaharu, 49
GCMS (General circulation models), 283
Gede-Pangrango National Park, 181
General Agreement on Tariffs and Trade (GATT), 243
General circulation models (GCMS), 283
Genetic improvement, 121–122
Genetic resources, 25–26, 121–122, 247, 253, 263
Gesneriaceae, 99
Ghana, 242
Gibbons, 72, 75–76, 78, 80, 211
Global climatic change, 264–269
Global warming, 34, 265–267, 278–279
Greenhouse effect, 34, 264–265, 279
Greenhouse gases, 265, 266–268, 280, 283
Gunung Api, 97–98
Gunung Behtuang dan Karimun Nature Reserve, 67
Gunung Binaia, 98
Gunung Gading, 71
Gunung Mulu National Park, 91, 131, 206, 211–212
Gunung Silam, 99

Habitats: forest, 32, 145–147, 176, 208; management of, 211; wildlife, 20, 30, 34–35, 54
Hardwoods, 12–14
Heat balance, 34
Heath forests, 9, 31, 89, 135
Heavy machinery, 106, 110, 112
Hedge orchards, 119
Herbaceous plants, 20, 99, 108, 208
Hong Kong, 229
Hornbills: conservation of, 135; densities of, 73–74; hunting of, *71*, 76–80, 82–83; on shifting-cultivation land, 72; as seed dispersers, 58, 83; surveys of, 71, 209
Hose Mountain National Park, 211
Huay Kha Khaeng-Thung-Yai Narusan Sanctuary, 164
Human impact, 60, 66, 137, 176, 265, 280. *See also* Hunting; Logging

Human needs, and biological diversity, 135
Human resources, and conservation, 134, 182–184, 209. *See also* Employment
Hunting: control of, 82–83, 209; and ecosystems, 82–83; effects on wildlife, 66, 76–80, 82–83; impact of, 71–72; levels of, 69–71, 77–81; photograph of, *69;* protection from, 31, 63, 208; and Totally Protected Areas (TPAs), 212–213; and tropical rainforests, 247; and vertebrate distribution, 60, 62
Hydrological balance, 34
Hydrology, 88, 91, 109, 112
Hydroxyl radical, 278

Import tariffs, 242–244
In vitro propagation, 120
India: mangrove forests of, 10; protected areas of, 144, 155; and rattans, 28; species richness of, 142, 181; tropical rainforests of, 5
Indigenous people, 137, 226–227. *See also* Customary rights; Native land rights; Penan tribe
Indochina, 28, 153
Indonesia: conservation in, 133, 162, 167; ecological studies in, 130–134, 137; endemic species of, 148; forest habitats of, 30; forest regeneration in, 133; geography of, 129–130; human resources of, 134; map of, *6;* and natural resource accounting, 252; nature reserves in, *136;* protected areas in, 157, 166; pulse harvesting in, 250; and rattan exports, 28–29; reserves in, 152–153; species richness of, 142, 148–149; tropical hardwood production, 13; tropical moist forests of, 143–147, 152; vegetation of, 129–130
Indo-Pacific region: biogeographic considerations of, 166–167; biological richness of, 148–155; conservation investments of, 159–161; conservation recommendations for, 162–166; investment priorities for, 163–164, 167–168; landscape patterns of, 155–158; lowland rainforests of, 140–141, 169–172; protected areas of, 173–174; protection status of, 146; regional conservation patterns, 143–147; threats to tropical forests of, 141–143; tropical moist forests of, 140–141, 169–172

Insects, 54, 117
International trade, 12–14, 245–246
Invertebrates, 142
Investment: and biological diversity conservation, 143, 158–162, 165–166; and Conservation Potential/Threat Index (CPTI), 140; long-run, 254; priorities for, 158, 162–164, 167–168; and timber concessions, 254
Irian Jaya, 10, 129–130, 146, 153–155, 157, 166
Ironwood tree, 57
Island biogeography, 176
Islands, as conservation sites, 180

Jambu, 55
Japan, 229, 269
Java, 23, 129–130, 145–416, 157, 166, 181
Javan rhinoceroses, 58
Jengka Forest Reserve, 22, 91, *249*

Kalimantan: ecological studies on, 130–131, 133; geography of, 129; protected areas of, 155, 157; species richness of, 21, 146; tree species of, 26
Kapur seedling, *110*
Kayan Mentarang Nature Reserve, 134–*135*
Kelumba Wildlife Reserve, 32
Kerangas forests: and biomass, 46–48; map of, *47;* nontimber resources, 49–50; photographs of, *44–46;* production of, 44–46; stock densities of, 45; sustainable management of, 48–50
Kerangas soils, 41
Kerapah forests: and biomass, 46–48; map of, *47;* production of, 44–46; sustainable management of, 48–50
Kerapah soils, 41
Keystone plants, 59, 62, 158
Kinabalu Park, 55, 180–181
Kuala Belalong, 71
Kubah National Park, 79, 207

La Selva Research Station, 92, 94, 131
Lambir Hills National Park, 93, 206–207, 209
Land use, 212, 222, 268–271, 274
Landscape analyses, 143, 167
Landscape patterns, 155–158, 211

Langurs, 72, 75–76, 78, 211
Lanjak-Entimau Wildlife Sanctuary, 67, 209–211
Laos, 155, 157, 162, 165
Latin America, 277
Lauraceae, 20, 27, 58, 59, 100
Leaf litter, 60
Leaf nutrient concentrations, 92
Lecythidaceae, 9
Legume family, 59
Leguminosae, 20
Lesser Sunda Island, 23, 157, 166
Liberation thinning, 116, 253
Limbang Mangroves Wildlife Sanctuary, 208
Limestone, 30, 34, *97–100*, 181
Line transect surveys, 70
Litterfall, 92, 108
Loagan Bunut National Park, 32, 209
Logging: and biological diversity, 247, 259; and carbon, 274; of dipterocarp forests, 80, *241;* effects on soil characteristics, 105–113; effects on wildlife, 81; and forest production, 96, 206; low-impact, 50; and mammals, 80; and native people, 233–234; and nutrients, 95–96; of old-growth forests, 247, 258; photographs of, *107;* pressures from, 158, 162, 182, 208; recovery from, 12, 83; selective, *68,* 105–106, *249, 275;* and soil erosion, 95, *107–109,* 113, 224, 234, 257, 264; studies of, 67–70, 75–77; sustainable, 49; and Totally Protected Areas (TPAs), 212; in tropical moist forests, 140; and vertebrate distribution, 60. *See also* Timber harvesting
Log-landing areas, 106–109, 111–113
Lowland rainforests: in biogeographic units, 150–152, 171–172; and breeding populations, 61–63; comparisons by country, 144; forest formations of, 9–10; in Indo-Pacific region, 140–141, 169–170; as protected areas, 153
Luzon, 147
Lysimetry, 88

Macaques, 73, 76, 78
Madagascar, 177
Magpie robins, 75, 81

Mainland conservation sites, 180
Malay Peninsula. *See* Peninsular Malaysia
Malaya. *See* Peninsular Malaysia
Malayan sun bear, 70
Malayan uniform system, 250, 256–257
Malaysia: deforestation rate of, 250; mammals in, 60; protected areas of, 30–31, 145, 165–166; and rattans, 28–29; species richness of, 142–143, 148; tropical moist forests of, 152, 158, 165; vertebrates in, 54
Malaysian tree squirrels, 56
Malesian tropical rainforests: conservation of, 29–36; distribution of, 5; ecology of, 20–22; economic importance of species in, 25–29; endemic species of, 21–22, 24–25; forest formations of, 9–10; phytogeographic relations of flora in, 22–25; plant diversity of, 19–24; protected areas of, 29–33; tree species of, 26. *See also* Southeast Asian tropical rainforests
Maliau Basin Conservation Area, 61–62
Maludam Wildlife Sanctuary, 208
Maluku, 146, 166
Mammals: conservation of, 135, 210; diversity of, 75; endemic species of, 149–150; food supply of, 59; hunting of, 76–77, 82; and logging, 80; on shifting-cultivation land, 72; and rainforest ecosystems, 54; as seed predators, 117; and soils, 60; species richness of, 142, 148–150; surveys of, 71; terrestrial, 57–58; and timber production, 62; and tree plantations, 62–63. *See also* Vertebrates; Wildlife; *names of specific mammals*
Mango, 55
Mangrove forests: conservation of, 32; in Indonesia, 131; protection of, 31; in Sarawak, 208; soils of, 89; in Southeast Asia, 10; tree density of, 22
Maracá Island, 89, 92–93
Medicinal plants, 25, 27, 33, 55, 176, 247
Meliaceae, 20, 28, 58, 59
Menchali Forest Reserve, 9
Menispermaceae, 20, 27
Methane, 265–266, 278–279
Micropropagation, 120–121
Mindanao, 147
Mineral content, 61, 94, 97–100

Minimum diameter cutting limits, 48, 69, 105, 223, 255–257
Moluccas, 23
Monocyclic uniform systems, 256
Monsoon forests, 5, 130
Montane forests, 32, 61, 93–94, 135
Moraceae, 20, 28, 58, 59
Mt. Bloomfield, 99–100
Mt. Kinabalu, 21, 32, 180–181
Mt. Lotung Area, 32
Mt. Silam, 91
Multiple-use management, 258
Mycorrhizas, 93–94, 111, 119
Myristicaceae, 8
Myrsinaceae, 55
Myrtaceae, 20, 99–100

Nanga Gaat, 67–68, *75–76*
National parks, 204–206, 214–218, 222. *See also names of specific parks*
Native land rights, 221, 224–226, 233, 238. *See also* Indigenous people
Natural forest management, 50
Natural resource accounting (NRA), 252
Nature reserves: distribution of, 154; in Indonesia, 152–153; in Sarawak, 204–206; size of, 158, 162, 209; transfrontier, 155, 157–158, 162. *See also names of specific reserves*
Net primary production (NPP), 44–45
New Caledonia, 99
NGOs (Nongovernment organizations), 31, 224
Niah National Park, 206
Nickel, 99–100
Nitrous oxide, 265–266, 278–279
Nondegraded soils, 110
Nongovernment organizations (NGOs), 31, 137, 224
Nontimber resources, 49–50
NPP (Net primary production), 44–45
NRA (Natural resources accounting), 252
Nursery facilities, 117–122, *120*
Nutrients: in biomass, 87–88; cycling of, 109; in leaves, 92; and logging, 95–96; measurement of, 88–89; in soil, 87–88, 109, 113; and tree plantations, 96

Oil palm, 236, 238
Oil palm plantation, *270*
Oils, 176
Old-growth tropical forests, 241–242,
 245–248, 258
Open soil systems, 88
Open-strip surveys, 70
Orangutans, 56–57, 59, 60–62, 81–82, 209
Orchids, 21, 27, 83, 209
Ornamental plants, 27–28, 33
Oxisols, 9

Palawan, 146–147, 149
Palms, 32, 209
Papua New Guinea: endemic species of,
 23–24, 148, 165; forest habitats of, 30; plant
 diversity in, 19, 21, 181; pulse harvesting in,
 250; reserves of, 155, 166; species richness
 of, 148; tropical hardwood production in,
 14; tropical moist forest of, 143–145
Pasoh Forest Reserve, 21–23, 28, 89, 131
Peatswamp forests: and biomass, 46–48;
 conservation of, 32; as forest formation,
 9–10; of Malesia, 9–*10;* nontimber
 resources, 49–50; phasic communities of,
 42–44, 48; photograph of, *49;* production
 of, 44–46; protection of, 31; in Sarawak, *47,*
 208–209; soils of, 89; sustainable management of, 48–50; timber harvesting in, 223;
 tree density of, 22, 45
Peatswamp soils, 42
Penan tribe, 79, 224–227, *225, 228,* 232–*233,* 238
Peninsular Malaysia, 11, 98, 166: biomass
 reduction of, 274; endemic species of,
 23–24; minimum-diameter cutting limits
 in, 255–257; plant diversity in, 22, 181; protected areas of, 31–32, 146; pulse harvesting
 in, 250; soils of, 89, 99; species in, 26, 32–33;
 studies of, 59; and sustainability, 252
Permanent Forest Estates (PFEs), 211, 213: and
 conservation, 62–64, 83; extension of, 63; as
 forest resource, 221–222; in Sarawak, 204,
 211, 213. *See also* Protected areas; Totally
 Protected Areas
Permanent-plot monitoring system, 135
Peru, 106
Petai, 55

Petrochemical industry, 237
PFEs. *See* Permanent Forest Estates
Phasic communities, 42–44, 48
Pheasants, 71, 73–74, 79–80
Phenology, 133
Phenotypic selection, 121–122
Philippines: dipterocarp trees in, 11; endemic
 species of, 23–24, 142, 148–149; forest habi-
 tats of, 30, 145–147; forest sector develop-
 ment in, 242; mangrove forests of, 10;
 protected areas of, 157–158, 165–166; pulse
 harvesting in, 250; and rattan exports, 29;
 tropical hardwood production of, 13; ultramafic rocks vegetation of, 99
Phytogeography, 22–25, 33
Pigeons, 58, 78
Pigs, 57, 60, *70,* 83
Pineapples, 238
Pioneer tree seedlings, 112–113
Plant diversity: conservation of, 29–36; conservation sites, 176–202; ecology of, 19–21,
 176–177; economic importance of, 25–29;
 identifying areas of, 177–178, 185; and phytogeographic relations, 22–25; in Sarawak,
 208; site analysis, 181–184; site selection,
 179–181; tropical Asian centers of, 186–202.
 See also Biological diversity; Species richness
Plantain squirrels, 55, 81
Plant-animal interactions, 54, 59–61, 69–70,
 83–84, 133
Plantations, 50, 62–63, 95–96, *249, 270*
Plants: aquatic, 20; economic products of,
 25–29, 33, 54–55; endemic species of, 149; ex
 situ conservation of, 208; flowering, 19–20,
 22; herbaceous, 20, 99, 108, 208; keystone,
 59, 62, 158; medicinal, 25, 27, 33, 55, 176, 247;
 ornamental, 27–28, 33; phytogeography,
 22–25, 33; uses of, 176; vascular, 19, 22, 25, 27,
 142, 176–177, 181. *See also* Plant diversity;
 Tree species; *names of specific plants*
Podzols, 41, 67
Pollination, 54–55, 63
Polycyclic selection systems, 256
Porcupines, 56–57, 60
Poring, 55
Poverty, 231–233

Precipitation, 284–285
Prices, 242–247, 254
Primary forests, 72–74, 81–83, 176
Primates: conservation of, 137; densities of, 72;
    food supply of, 59; hunting of, 70, 76–78,
    82–83; and logging, 75; on shifting-cultiva-
    tion land, 72; populations of, 212; as seed
    dispersers, 56, 82; surveys on, 209. *See also*
    names of specific primates
Proboscis monkeys, 60, 62, 209, 212
Process, and biological diversity, 135
Program research, 131–133
Project-by-project research, 131–133
Propagation, 118–123
Protected areas: conservation of, 144; financ-
    ing mechanisms for, 168; forest resources
    of, 141; isolation of, 157, 167; landscape
    analyses of, 143; landscape patterns of,
    155–158; legal protection of, 182; legal status
    of, 144; management of, 158, 162, 165,
    182–184; networks of, 165, 168, 182; size
    distribution of, 151–155; spatial distribution
    of, 141, 155–158, 173–174. *See also* National
    parks; Nature reserves; Permanent Forest
    Estates; Totally Protected Areas; Wildlife
    sanctuaries
Pulau Tukong Ara-Banun Wildlife
    Sanctuary, 210
Pulong Tau National Park, 79, 207, 211
Pulse harvesting, 250

Queensland, 5, 89

Rail babbler, 75
Rainforests. *See* African tropical rainforests;
    American tropical rainforests; Southeast
    Asian tropical rainforests; Tropical rain-
    forests; *names of specific types of forests*
Ramin forests, 42, 45, 48
Ramin trees, 10, 223
Raptors, 83
Rats, 60
Rattans: in dipterocarp forests, 20; in keran-
    gas forests, 49; and Penan tribe, 227; in
    Southeast Asian rainforests, 9, 25, 28–29, 33
Recreational opportunities, 205, 247. *See also*
    Tourism

Reforestation, 280–281. *See also* Forest regen-
    eration
Reptiles, 54, 142
Research, 131–137. *See also* Ecological studies
Reserves. *See* Nature reserves
Rhinoceroses, 58, 61, 82, 207, 209
Rio Palenque, 21
Rodents, 56, 63
Rompin-Endau forest, 31
Rooting systems, 123
Roundwood, 246–247
Rubiaceae, 20, 27
Rural population, 66, 221, 232–233, 235, 238
Rutaceae, 27

Sabah: and breeding populations, 63; diptero-
    carp trees of, 11; endemic species of, 21, 148;
    forest regeneration in, 116; logging in,
    105–113; plant diversity in, 21; plant-animal
    interactions in, 59–61; protected areas in,
    32, 146, 155; pulse harvesting in, 250; and
    sustainability, 252; tree species of, 26; tropi-
    cal hardwood production in, 13–14
Sabal Forest Reserve, 46, *46*
Sago, 238
Salt licks, 61, 76
Sambar deer, 80
Samunsam Wildlife Sanctuary, *44*, 206, 209,
    212
San Carlos de Rio Negro, 46
Sapindaceae, 20, 28, 55, 59
Sapotaceae, 8, 20
Sarawak: biological diversity in, 204–218; and
    biomass, 46–48; and conservation, 32;
    ecological research of, *132;* economy of, 227,
    229–230, 233–238; employment in, 230–231;
    endemic species of, 21, 148; forestry policies
    of, 221–224; gross domestic product of, 229;
    growth studies of, 93; kerangas forests of,
    44–47; land use in, 222; national parks in,
    204–206, 214–218; native tribes of, 224–227;
    natural forest cover in, 223; nature reserves
    in, 204–206; oil palm plantation of, *270;*
    plant diversity of, 22; political environment
    of, 224; population of, 226–227; poverty
    level of, 231–233; production and produc-
    tivity of, 44–46; protected areas in, 32–33,

Sarawak *(continued)*
146, 155; pulse harvesting in, 250; soils of, 41–42, 99; study areas of, *67–70*; and sustainability, 252; and sustainable management, 48–50; timber industry in, 221–239; Totally Protected Areas (TPAs) of, 204–218; tree species of, 26; tropical hardwood production in, 13–14; vegetation of, 42–44; wildlife of, 66–84; wildlife sanctuaries in, 32, 204–206, 214–218

Sarawak Mangroves National Park, 208

Scyphostegiaceae, 23

Second-growth forests, 241–242, 253–258

Sedimentary rocks, 99

Seed dispersal, 55–59, 82–83

Seed predators, 56, 82, 117

Seladang, 58

Selection-silvicultural system, 48

Selective logging, *68*, 105–106, *249*, 275

Sempilor, 49

Shifting-cultivation land, *68*, 72–74, 81, 83, 272, 280

Silviculture, 12, 48, 50, 116, 256. *See also* Forest management

Simaroubaceae, 27

Singapore, 10

Sinharaja forest, 165

Skid trails, 106–109, 111–113

Slash-and-burn agriculture, 48, 105, 182

SLFs (State land forests), 222

Soil bulk density, 106

Soil compaction, 106–108, 112

Soil erosion, and logging, 95, *107*–109, 113, 224, 234, 257, 264

Soil water, nutrient concentrations in, 88, 91, 109, 112

Soils: alluvial, 60; analyses of, 87–88; and biomass reduction, 275; and carbon, 34, 273; chemical properties of, 107; and climate, 89; deep, 88; degraded, 110; differences in, 91–92; of dipterocarp forests, 89; fertility of, 105; and forest management, 50–51, 87–100; and forest regeneration, 110–112; and forest stature, 92–93; and forest type, 89–92; and hydrology, 88, 91, 109, 112; influence of, 89; kerangas, 41; kerapah, 41;

logging effects on, 105–113; and mammal distribution, 60; mineral content of, 61, 94, 97–100; and mycorrhizas, 93–94; nondegraded, 110; nutrient-deficient tropical soils, 41–42, 88; nutrients in, 87–88, 113; oxisols, 9; peatswamp, 42; physical properties of, 107; podzols, 41, 67; shallow, 88; spodosols, 9, 41; temperature of, 284; ultisols, 9

Solomon Islands, 143, 145, 165

South Korea, 229

Southeast Asian tropical rainforests: and Centers for Plant Diversity (CPD) project sites, *183*, 186–202; climate of, 5–7, 285; conservation site selection, 180–181; conservation status of, 184; deforestation rates, 272–275, 277–278; distribution of, 5–10; floristic composition of, 8–9; forest formations in, 9–10; and hardwood trade, 12–14; map of, *6;* seed dispersal in, 54–60; species richness of, 7–8; threat to, 184; timber harvesting in, 105; vertebrates of, 54–64. *See also* Dipterocarp forests; Malesian tropical rainforests; *other specific types of forests*

Species: colonizer, 81; economic importance of, 25–29; edge, 81; endangered, 207, 209, 211–212; extinction of, 25, 176; habitats of, 208; timber, 12–14, 26–27, 33, 208, 255; types of, 88–92. *See also* Endemic species; Tree species

Species diversity. *See* Biological diversity; Plant diversity

Species richness: of Indo-Pacific region, 148–155; of lowland rainforests, 140–141; in Sarawak, 208; and site selection, 179–181; in Thailand, 162; of tropical moist forests, 142–143; of tropical rainforests, 7–8. *See also* Biological diversity; Plant diversity

Spiderhunters, 54

Spodosols, 9, 41

Squirrels: densities of, 73–74; food supply of, 59; hunting of, 77–79; and logging, 75–77; on shifting-cultivation land, 72; as seed dispersers, 55–56; as seed predators, 82

Sri Lanka, 5, 144–145, 148, 162, 164–166

State land forests (SLFs), 222

Stranglers, 20

Stumpage values, 243, 245, 251–255, 258

Stunted forests, 89–90

Sulawesi, 11, 129–130, 146, 153, 155, 166

Sumatra: dipterocarp trees in, 11; ecological studies of, 131; endemic species of, 23; geography of, 129–130; protected areas of, 146, 153, 155, 157, 166

Sumatran rhinoceroses, 82, 207, 209

Sunbirds, 54

Sunderbans forest, 10

Sungai Beloh, 72

Suriname, 96, 106

Sustainability: and biomass, 46–48; definitions of, 248–253, 259; even-flow, 250–251, 259; management of, 48–50, 129–138, 222–223, 239, 239; of timber production, 247–253. See also Forest management

Swamp forests, 32

Swiftlets, 209

Symplocaceae, 99

Tabin Wildlife Reserve, 32

Taiwan, 144, 151, 157, 166, 229

Taman Negara, 24, 31–33

Tanjung Datu, 71

Tanjung Puting Reserve, 56

Tannins, 55, 176

Tapirs, 58, 82

Tawan Hill, 32

Technical forest management, 133

Temperate timber products, 243, 245, 247, 254, 258

Terns, 209

Terrestrial herbivores, 63

Terrestrial mammals, 57–58

Thailand: boom-and-bust pattern, 242; and fire, 63; protected areas of, 145, 152, 155, 158, 164; and rattan exports, 28–29; species richness of, 162

Threatened Plants Unit (TPU), 179

Thymeleaceae, 99

Tigers, 61

Timber harvesting: and conservation, 223–224; cutting cycles of, 223; of old-growth forests, 246; policies concerning, 222–223;

prohibition of, 32, 211; rate of, 245; in Southeast Asian rainforests, 105; and sustainable management, 48–50, 209, 221; and vertebrates, 62. See also Logging

Timber industry: boom-and-bust pattern of, 241–244, 259; economic contributions of, 26–27, 33, 55, 227, 229–230, 237; effects on wildlife, 66; employment in, 230–231; and native land rights, 224–226; and prices, 245–247; problems in, 233–234; products of, 236–237, 242, 245–246; role of, 235–236; sawmills, 231; and second-growth forests, 253–258; social impacts of, 231–233; and stumpage values, 243, 245, 251–255, 258; sustainable production, 9, 247–253, 259; and timber concessions, 253–255; and timber depletion, 244–247; and timber species, 12–14, 26–27, 33, 208, 255. See also Forest management

Tissue culture techniques, 120

Topography, 91

Topsoil, 60, 109

Totally Protected Areas (TPAs): and biological diversity, 207–208; as forest resource, 221; forest types in, 205; roles of, 206–207; in Sarawak, 204–218; size of, 211, 213; system of, 83–84. See also Permanent Forest Estates (PFEs); Protected areas

Totally protected forests, 31

Tourism, 168, 182, 206, 236–237, 247

TPAs. See Totally Protected Areas (TPAs)

Trade, international, 12–14, 245–246

Transfrontier reserves, 155, 157–158, 162, 210

Translocation, of vertebrates, 62–64

Tree plantations, 62–63, 95–96, 249

Tree species: economic importance of, 25–27; endemic, 32; as food supply, 62; in Malesian tropical rainforests, 21; as nesting sites, 60, 62; propagation of, 118–122; and species richness, 7; substitution among, 255; timber trees, 12–14, 26–27, 33, 208, 255; and tree density, 22, 45; of tropical rainforests, 8–9. See also Forest management; Logging; Timber harvesting; Timber industry; specific types of forests

Treeshrews, 55, 75–76

Tropical moist forests: in biogeographic units, 149–151, 166–167, 171–172; biological diversity of, 142–143, 158, 162; comparisons by country, 144; conservation of, 140–141, 159–168; as protected areas, 153; regional patterns of, 143–147, 158; size distribution of, 151–155, 169–170; spatial distribution of, 155–158, 173–174; species richness of, 148–149; threats to, 141–143. *See also* Lowland rainforests; *other specific types of forests*

Tropical rainforests: and carbon management, 280–282; climate of, 5–7, *6*, 263–264, 285–286; comparisons of, 5–14; customary ownership of, 165; distribution of, 5–10; ecological roles of, 19; ecosystems of, 54–64; evergreen, 5, 7; and fertilization experiments, 94–95; floristic composition of, 8–9; forest formations of, 9–11; seasonal, 5; semi-evergreen, 5, 7; soils of, 9; species richness of, 7–8; topography of, 91; types of, 88–92. *See also* African tropical rainforests; American tropical rainforests; Southeast Asian tropical rainforests; *specific types of forests*

Ultisols, 9
Ultramafic rock formations: natural vegetation on, 30, 32, 34–35, 181; rainforests on, 87, *97*, *99*–100; and stunted forests, 89
Ulu Segama Forest Reserve, 105–106, 108, 117, 122
Undisturbed forest, 106–108
Ungulates, 59, 70, 76, 82
United States, 269
Urban development, 212
Usun Apau National Park, 211

Vascular plants: extinction of, 176–177; medicinal, 27; in Sabah, 22, 181; in Southeast Asia, 19, 25; species richness of, 142
Vegetation: carbon in, 273; of Indonesia, 129–130; and limestone, 98–99; of Sarawak, 42–44; selective logging damage, 105;

studies of, 133; types of, 89; and vertebrates, 59. *See also* Plant diversity; Plants
Vegetative propagation, *118–121*
Vertebrates: conservation of, 30, 54–64, 135; distribution of, 60–62; diversity of, 82; and forest ecosystems, 82; pollination by, 54–55, 63; seed dispersal by, 55–59, 63; studies of, 177; translocation of, 62–64. *See also* Wildlife
Vietnam, 148, 155, 157, 165–166
Vines, 110–112
Vitaceae, 20
Vochysiaceae, 9, 12

Water: flows of, 263; nutrients in, 88, 96; and soil, 88, 91, 109, 112; and timber harvesting, 247, 284; and vertebrate distribution, 61
Water catchments, 207, 234
Watersheds, 88, 224, 252–253
Western Europe, 269
Wetlands, 130
Wild cattle, 58
Wild pigs, 57, 60, 70
Wildings, 117–118, 120, 123
Wildlife: abundance of, 70; breeding of, 61–64, 81–82; and ecology, 133; endangered, 212; ex situ conservation of, 208; habitats for, 20; hunting of, 66, 76–80, 82–83; and logging, 75–77, 80–81; management of, 83–84, 209; and primary forests, 72–74, 81–83; rehabilitation centers, 222; and shifting cultivation, 72–74, 81; study areas, 67–70; study methods, 70–72. *See also* Birds; Mammals; Plant-animal interactions; Vertebrates; *names of specific wildlife*
Wildlife sanctuaries, 31–32, 204–206, 214–218, 272. *See also names of specific sanctuaries*
Wood fuels, 55, 176, 271, 281–282
Wood products, 237, 243, 247, 255. *See also* Timber industry

Yanamomo, 7–8

Zingiberaceae, 27